ASTRONOMY AND ASTROPHYSICS LIBRARY

Springer

Berlin
Heidelberg
New York
Barcelona
Hong Kong
London
Milan
Paris
Tokyo

Physics and Astronomy

ONLINE LIBRARY

http://www.springer.de/phys/

Wolfgang Kundt

Astrophysics

A Primer

With 40 Figures and 14 Color Plates

 Springer

Professor Dr. Wolfgang Kundt

Institut für Astrophysik
Universität Bonn
Auf dem Hügel 71
53121 Bonn, Germany
E-mail: wkundt@astro.uni-bonn.de

Cover picture: The radio galaxy 0756 + 272 of (Eilek) type B, or 'tailed source', whose jet heads – or stagnation surfaces – are thought to no longer plough supersonically into their ambient medium. The galaxy belongs to Abell 610. It has a redshift of z=0.0956; its radio lobes have an angular diameter of 160". Its jet substance is thought to be pair plasma performing an $\vec{E} \times \vec{B}$-drift. The AGN is some 0.7 cm to the right of the bright (red-cored) central hotspot. [Image courtesy of Frazer Owen, NRAO/AUI]

Library of Congress Cataloging-in-Publication Data applied for.

Die Deutsche Bibliothek – Cip Einheitsaufnahme

Kundt, Wolfgang:
Astrophysics: a primer / Wolfgang Kundt. – Berlin; Heidelberg;
New York; Barcelona; Hong Kong; London; Milan; Paris; Singapore;
Tokyo: Springer, 2001
(Astronomy and astrophysics library)
(Physics and astronomy online library)
ISBN 3-540-41748-6

ISSN 0941-7834
ISBN 3-540-41748-6 Springer-Verlag Berlin Heidelberg New York

Springer-Verlag Berlin Heidelberg New York
a member of BertelsmannSpringer Science+Business Media GmbH

http://www.springer.de

© Springer-Verlag Berlin Heidelberg 2001
Printed in Germany

Typesetting: Data conversion by LE-TEX, Leipzig
Cover design: *design & production* GmbH, Heidelberg

Printed on acid-free paper   SPIN: 10831348   55/3141/tr - 5 4 3 2 1 0

Preface

This book is based on a class that I lectured repeatedly during the years 1986 through 1999, at Boston, Bangalore, Bonn, Siegen, and Linz. Occasionally I lectured it within two or more terms. In each case, it was my endeavour to provide my audience with as complete as possible an overview of the physical methods and results used in present-day *astrophysics*, in order to gain an insight into the functioning of the Universe.

Here, the true climax of cosmic insight appears to me to be the *Anthropic Principle*, whose strongest variant reads: The state parameters and laws of physics are such that after a suitable interval of time – of the order of 10 Gyr (Gyr= 10^9 yr =: 1 geon) – a biosystem can evolve on a suitable planet or moon of a suitable star or planet, with e.g. *Homo Sapiens* as a sensing observer. I.e. the functioning of our animated (!) planet Earth cannot only be described in physical terms, but its existence dictates, indirectly, the structure of the laws of physics. Thus, you and I would not exist, for example, if the mean cosmic substratum was (only) some 10^2-times denser than it is: The Universe would have recollapsed before higher terrestrial life could evolve. The biological clock is tuned both to the solar clock, and to the cosmic clock.

But Earth does not seem to take an outstanding part in the Universe; it obeys the same laws. Time and again, the *Copernican Principle* can serve as a useful guide: We live on a typical planet which moves around a typical star inside a typical (spiral) galaxy inside a typical cluster of galaxies, and which has been animated for a typical cosmic era. Admittedly, we play a considerable role in what happens in the near vicinity of Earth – by our technical achievements and mental insight – but we do not really play a role that would upset the Universe. We are rare, but not central, and not forbidden.

Astrophysics is not exactly a subfield of physics, but an extension of physics: It concerns itself with material objects on much larger (than terrestrial) length scales, whose substratum has, in general, much lower densities, and exerts much lower pressures, though occasionally also much higher ones, whose temperatures, magnetic fields, electric voltages, and velocities have much larger ranges than on Earth. Often, we have to extrapolate our knowledge. Such extrapolations involve natural uncertainties, as do the precise local circumstances of an observed object which, in general, cannot be retrieved reliably from its observations. We must therefore be careful with our interpretations, more careful than *textbook knowledge* occasionally is: subjunctive moods cannot be avoided by a trustworthy description.

I shall stress such possible weak spots in our insight, candidates for the subjunctive mood, wherever they occur to me, in order to help prevent research from getting stuck in dead ends. The existence of such weak spots is shared with all exact sciences whose theories are not immediately testable – like geophysics – whenever processes run in the Earth's unaccessibly deep interior, or unnoticeably slowly, as in volcanism and plate tectonics – or biophysics – whenever objects of interest are submicroscopically small so that their perturbation-free recording in vivo is difficult, such as the mono-cellular water pumps in the endodermal walls of root tips. They have often led to violent, loud or mute controversies (between individuals and/or schools), without any real necessity: The evolution of insight would probably fare better without scientific revolutions if *alternative interpretations* were considered legitimate, and were always mentioned, as has been repeatedly emphasized, e.g. by Richard Feynman.

This book summarizes a significant fraction of my life's work, and owes its existence to decades of weekly seminars and talks whose participants and speakers helped in shaping my insight. Among them are my Hamburg 'master' Pascual Jordan and co-students Jürgen Ehlers and Klaus Hasselmann, my Bonn 'patron' Wolf Priester, as well as my successive associates, students and friends Hans Heintzmann, Eckhard Krotscheck, Max Camenzind, Hajo Leschke, Marko Robnik, Axel Jessner, Ashok Singal, Reinhold Schaaf, Daniel Fischer, Hsiang-Kuang Chang, Carsten van de Bruck, Hans Baumann, and Gernot Thuma. Stefan Wagner helped with state-of-the-art spectra. Reinhard Schlickeiser triggered off the introductory lectures which have materialized in this book. To all of them go my hearty thanks. During the 18 months of preparation and work on this book, Markus Draxler introduced me knowingly and patiently to the use of the routine *Scientific WorkPlace,* and Günter Lay and Horst Scherer followed him up masterfully. Stefan Siegel converted the numerous paper drawings into printable eps files, and Vadim Volkov and Martin Hetzer proofread the chapter on Astrobiology. Again, my thanks to all of them!

Bonn, July 2001   *Wolfgang Kundt*

Contents

Useful Numbers

$\alpha \;\; := e^2/\hbar c$	fine-structure constant	$1/137 = 10^{-2.1368}$
$\alpha_G := Gm_p^2/\hbar c$	gravitational fine-structure constant	$10^{-38.23}$
AU	astronomical unit	$10^{13.175}$ cm
c	speed of light in vacuum	$10^{10.4768}$ cm/s
e	elementary charge	$10^{-9.3185}$ esu
eV	electron volt	$10^{-11.7954}$ erg
G	gravitational constant	$10^{-7.176}$ dyn cm^2/g^2
$\hbar := h/2\pi$	rationalized Planck's constant	$10^{-26.9769}$ erg s
k	Boltzmann's constant	$10^{15.8599}$ erg/K
lyr	light year	$10^{17.9759}$ cm
$L_\odot$	solar luminosity	$10^{33.5828}$ erg/s
m_e	electron rest mass	$10^{-27.0405}$ g
m_p	proton rest mass	$10^{-23.7766}$ g
$M_\odot$	solar mass	$10^{33.2986}$ g
$M_\oplus$	Earth's mass	$10^{27.7765}$ g
pc	parsec	$10^{18.4893}$ cm
$r_e := e^2/m_e c^2$	classical electron radius	$10^{-12.550}$ cm
$R_\odot$	solar radius	$10^{10.843}$ cm
$R_\oplus$	Earth's radius	$10^{8.804}$ cm
$S_\odot$	solar flux	$10^{6.1314}$ erg/cm^2 s
$\sigma_{SB} = (\pi^2/60)k^4/c^2\hbar^3$	Stefan–Boltzmann constant	$10^{-4.246}$ erg/cm^2 s K^4
$\sigma_T = (8\pi/3)(e^2/m_e c^2)^2$	Thomson cross section	$10^{-24.177}$ cm^2
$T_\odot$	solar temperature	$10^{3.76}$ K

$\log 2 = 0.30103$
$\log 3 = 0.47712$
$\log 5 = 0.69897$
$\log e = 0.43429$
$\log \pi = 0.49715$

1. Cosmic Structures

In this first chapter, the *cosmic players* (objects) are introduced superficially, whose detailed properties will then entertain us to the end of the book. The play starts in front of our doors, with the Solar System, and continues to our host galaxy, the Milky Way, and further to groups and clusters of galaxies until the outer reaches of observation, the Early Universe. Emphasis will be placed on length scales, time scales, masses, velocities, densities, temperatures, pressures, magnetic fields, and radiations. The newcomer can find more facts and observational backup, e.g., in [Karttunen et al., 2000].

1.1 Calculus and Notation

Astrophysics differs from other branches of physics by its difficult accessibility. A deficit of tangible and visible detail must be replaced by a surplus of imagination and cross-linkages, and *interpretations* require many more order-of-magnitude estimates than otherwise for which precision is less important than speed and transparency. Success in research thus depends more critically on a handy calculus.

As in every new field of physics, familiarising with the relevant *scales* is of paramount importance. For instance, civilized people are quite familiar with velocities of order several km/h or with temperatures of the order of 30°C, because we have memorised typical hiking, cycling, and driving speeds, and typical temperatures like freezing, and evaporating of water, room temperature, and body temperature. For me, the difficulties start already when I am asked for my temperature in units of Fahrenheit: switching to another system of units requires memorising the (usually linear) conversion rules, and is, in principle, avoidable. One system of units suffices for all of physics.

As such a preferred system, this book will use the *Gaussian cgs system*, because it avoids the conversion factors ϵ_0 and μ_0 accompanying electric and magnetic field strengths. Is there an objectively preferred system of units? The answer is "no": in principle, there are as many physical dimensions as different physical measurement setups, i.e. very many. But as most of them measure equivalent sets of numbers – as a consequence of laws of nature – they have been eliminated, and expressed in terms of earlier ones. Every law of nature permits the elimination of one dimension, until, in the end, all dimensions are again removed, in the *natural system* of units which involves the universal constants $c, G,$ and $\hbar$. In this natural system, the fundamental quantities length, time, mass, (rest) energy are expressed in terms of those of a Planck particle, of mass $M = \sqrt{\hbar c/G} = 10^{-4.7}$g , and (Compton wave-) length $10^{-34.8}$m. But such a dimensionless description of nature would have disadvantages: Not only would we have to memorise rather unfamiliar (large or small) numbers; we would, above all, lose the most useful dimensional test which greatly facilitates detecting errors, and which even helps in finding new formulae. For these reasons, cgs units have carried through, by realising a useful compromise.

Calculations can be done in awkward and non-awkward ways. A clever way is used by arithmetic artists who perform difficult calculations by heart: they multiply *logarithmically*, and thereby deal with fewer figures. Example: the fourth root of 390 625 equals $10^{5.6/4} = 10^{1.4} = 25$, whereby $\log 4 = 0.6$ and $\log 2.5 = 0.4$ have been used, at an accuracy of a few per cent. Higher accuracies can be obtained by using a calculator.

Logarithmic calculations flourished during the era of the slide-rule, and of the tables of natural and decadic logarithms. They suggest the following *index notation* of quantities and formulae, which is widely used but hardly ever systematically so:

$$A_x := A/10^x \dim(A) , \qquad (1.1)$$

where A is a physical quantity of dimension $\dim(A)$. Example: $v_7 = v/10^7 \text{cm s}^{-1}$; i.e. v_7 is a dimensionless number, viz. the velocity v expressed in units of $10^7 \text{cm s}^{-1} = 10^2$km/s, (which occurs frequently in the Universe).

As an advanced example of the conciseness and usefulness of the index calculus, let us anticipate formula (5.5) for the temperature dependence of the electric conductivity σ of a sufficiently dense hydrogen plasma: $\sigma_{14} = T_4^{3/2}$; in words: σ amounts to 10^{14}s^{-1} for a plasma temperature T of 10^4K, and grows with T as $T^{3/2}$. Another example is the connection between temperature and sound speed for a hydrogen gas, (1.10): $c_8 = \sqrt{T_8}$.

Powers of ten are alternatively expressed by prefixes, like {D, H, K, M, G, T, P, E} for 10^n with $n = \{1, 2, 3, 6, 9, 12, 15, 18\}$, and {d, c, m, μ, n, p, f, a} for 10^{-n}. Unfortunately, the AIP Style Manual deviates from above by using "d, h, k" instead for the first three enlargement prefixes, and "da" for "deka"; Springer Company has urged me to follow this ugly convention in the present edition.

1.2 Solar System

Let us start our excursion around the Universe at our front door, so to speak, with our Solar System. Our planet Earth has a radius $R_\oplus = 10^{3.8}$km $= 10^{8.8}$cm. It encircles the central star, the Sun – of spectral type G2, mass

$M_\odot = 10^{33.3}$g – at a distance of one *astronomical unit* AU $= 10^{13.17}$cm or 8 light minutes, at the Keplerian speed

$$v_\oplus = \sqrt{GM/r} = 10^{(-7.2+33.3-13.2)/2}\text{cm/s} = 10^{6.5}\text{cm/s} = 30 \text{ km/s} , \quad (1.2)$$

a speed which is known to correspond to a revolution period of one year $= 10^{7.5}$s. As a consequence of this motion, all celestial bodies (planets, nearby stars) perform small ellipses on the celestial sphere. The almost constant solar distance ($d =$ AU) implies that we receive from the Sun an almost constant energy flux (= radiation power per area) $S_\odot$, given by

$$S_\odot = L_\odot/4\pi d^2 = 10^{33.6-1.1-2\times13.2}\text{erg/cm}^2\text{s} = 10^{6.1}\text{erg/cm}^2\text{s} , \quad (1.3)$$

or $S_\odot = 10^{3.1}$W/m^2, the so-called *solar constant*; of which we have known since the late 1990s that it partakes in the 11-year-period solar magnetic oscillation (or rather: (22.2 ± 2)-yr period), with an amplitude of 0.1%, in phase with the sunspot number and with various other solar properties. (A reduction of radiation power in the spots is overcompensated by an increase from around them.) Here, the solar luminosity $L_\odot$ has been evaluated from

$$L_\odot = 4\pi R_\odot^2 \sigma_{SB} T_\odot^4 = 10^{1.1+2\times10.8-4.2+4\times3.76}\text{erg/s} = 10^{33.6}\text{erg/s} . \quad (1.4)$$

The Sun formed 4.6 Gyr ago, more or less simultaneously with its planets. We think that in this process, an angular-momentum excess of the contracting gas cloud caused the transient formation of a flat, proto-planetary disk whose particles revolved differentially and thereby exerted shear forces onto each other such that they gradually spiralled inward toward the disk's core, the forming Sun. In more detail, gas, dust, and larger condensations will have separated from each other, condensed and evaporated, and spiralled inward at different rates because of varying amounts of pressure support. The proto-planetary disk acted like a grand ultra-centrifuge and thereby provided the conditions for the formation of *planets* and their *moons*, with their different chemical compositions which show a monotonic dependence (e.g. of evaporation temperature) on solar distance. (Whereas the Universe consists primarily of hydrogen, the mantle of Earth consists primarily of silicon and oxygen, at comparable weights.) At the disk's center formed the young Sun, initially with the (minimal) rotation period of 3.6 h.

Such a high initial spin of the Sun appears to conflict with a backward extrapolation of its present spin period, of (27.3 ± 0.5)d, if evaluated from the present braking rate, the latter estimated from its mass loss $\dot{M} = -10^{-14}$M$_\odot$/yr via the *solar wind*. Uncertain in this estimate is the effective lever arm out to which the Sun forces its wind into (rigid) corotation: If that lever arm was as large as $30R_\odot$, ($R_\odot = 10^{10.8}$cm) – as repeatedly claimed by Lotova [1988] – i.e. some 10^2 times the solar inertia radius ($\approx R_\odot/3$), the present spindown rate would be as large as $\dot{J}_\odot = -10^{-10}J_\odot$/yr, and a large spin at birth would no longer appear implausible. Independently of this rather uncertain estimate, a formation of the Sun at the center of its accretion disk argues in favour of a maximal initial angular momentum, because gravitational and centrifugal forces should have balanced at its rotational equator, as on a Keplerian orbit.

A further, independent hint at a high initial spin of the Sun comes from the insight of the 1980s that, quite likely all newborn stars pass through the *bipolar-flow* stage – whereby *all* is maintained with the uncertainty inherent in every statistical statement – during which a star blows two antipodal supersonic jets into its circumstellar medium, parsecs long, at right angles to its accretion disk. The precise mechanism of blowing has remained controversial until today, but most authors make strong magnetic fields and high rotational velocities (of disk and/or star) responsible for its functioning; see Chap. 11. The bipolar-flow stage lasts some $10^{4.5}$yr, estimated from the age of the oldest stellar jet sources, and brakes the central (pre T-Tauri) star. The bipolar-flow energy is limited by the star's initial rotational energy.

Let us return to the Sun which, in its core, has burnt hydrogen to helium for 4.6 Gyr at an almost constant though slightly increasing burning rate (due to an increasing mass density there), as a so-called *main sequence* star, and radiates the thus-liberated nuclear energy at its surface at a (photospheric) temperature of

$$T_\odot = 5.77 \text{ kK} = 10^{3.76}\text{K} \tag{1.5}$$

into space, see Plate 1. At the same time, it blows the solar wind, first inferred from the existence of (two branching) cometary tails, and later from magnetospheric storms which succeed visible eruptions on the surface within $\gtrsim$two days. The solar wind blows unsteadily, at velocities between $10^{2.5}$km/s and $10^{3.2}$km/s, usually between 4×10^2 und 8×10^2km/s, whereby the lower values prevail at equatorial latitudes, the higher values at polar ones. It presses against, or confines the *interstellar medium* (ISM) of the Milky Way, out to a distance of at least $10^{8.3}$km $= 10^{2.1}$AU which has not yet been reached by the American Pioneer and Voyager spaceships (launched during the 1970s, and escaping at speeds, after swing-by, of 3.5 AU/yr). The solar mass loss dominates (presently) at low solar latitudes.

The position and structure of the edge of the *heliosphere* – the inner bowshock towards the ISM, or *termination* shock, and the heliopause, or *stagnation* surface – depend on the composition of the ISM and on whether its relative velocity, of some 25 km/s, means super- or subsonic motion; see Sect. 2.3. Conservatively, this ISM is conceived of as *warm* hydrogen, of (ionization) temperature 10^4K. Instead, it may consist of relativistic electrons and positrons, so-called *pair plasma*, which should be generated abundantly in coronal magnetic-field reconnections around compact and normal stars and which has revealed its omni-presence in the Milky Way through the (mapped!) 511-keV annihilation radiation and, indirectly, through a missing factor of 5 of sufficient warm hydrogen [Reynolds, 1990; eV $= 10^{-11.8}$erg].

The heliopause would be open in the first (supersonic, stretched) case but closed (ellipsoidal) in the second case. Whatever this tenuous medium, the solar wind screens its planets against it, also against the soft tail of the so-called *cosmic rays*, a highly relativistic plasma of probably Galactic origin, see Chap. 10.

Earth screens itself against the solar wind by its *magnetosphere*, of radius $\gtrsim 10R_{\oplus}$, which diverts it (because of its high electric conductivity). The *magnetotail* of Earth reaches far beyond the lunar orbit; it scans the moon at monthly intervals. Earth is probably the only animated planet of the Solar System because it is the only planet whose typical (surface) temperatures have always ranged between $0°C$ and $100°C$, i.e. allowed for liquid water. Note that a once-frozen Earth would hardly have thawed again, because of the high albedo of ice and snow, and correspondingly a once-dried-up Earth would never have formed rivers, lakes, and oceans again, because of its low albedo: thermal Earth has managed to pass between both Skylla and Charybdis. Its solar distance realizes the optimal distance, within a few per cent of an always wet planet, on which life appears to depend [Rampino and Caldeira, 1994].

As is well known, the Solar System contains eight or more additional planets whose solar distances obey, approximately, the rule of *Titius* and *Bode*: $d_n/\text{AU} = 0.4 + 0.3 \times 2^n$, $n = -\infty, 0, 1, 2, \dots$, where $n = 3$ counts the position of the asteroid belt (or gap) between Mars and Jupiter. It is presently unknown whether or not other planetary systems obey the same rule; those detected tend to contain higher-than-Jupiter masses on inner-planet orbits. Of general validity may be the approximate logarithmic equi-distribution, characterized by the factor 2^n. Note that Titius–Bode's rule does not make a prediction about the expected masses, and chemical compositions: Whereas the outer planets, starting with Jupiter, have near-solar compositions, the inner ones are strongly enriched with (above all) stony and carbonaceous material as well as iron alloys. Only in this way does Homo Sapiens find the ≈ 40 different chemical elements required for his functioning, see Fig. 1.1.

Besides the planets, the Solar System lodges their moons which may have formed in a similar way from orbiting accretion disks left over after the planets had formed from the main disk, by (first) chemical and (then) gravitational coagulation. This convincing mechanism has been cast into doubt recently for 'our' own moon because backward integration of its tidal interaction with Earth leads to a collision with it in the near past. This backward integration ignores geological evidence which argues in favour of an angular-momentum loss of the *Earth–Moon system* in the past, e.g. via friction on a then stronger solar wind (of a then stronger magnetosphere of Earth); it also ignores the difficulty of ejecting mass from a celestial body without subsequently re-absorbing it, after a certain number of intersecting orbits. The Solar System is an inexhaustible playground for physical exercises!

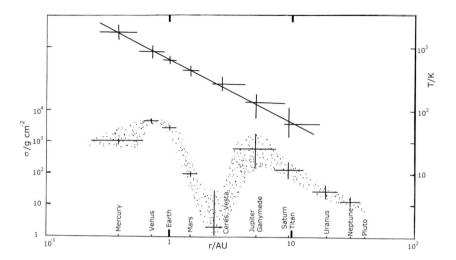

Fig. 1.1. Sketch of the inferred smoothed minimal mass distribution per area $\sigma(r)$ in the Solar System, and of the evaporation temperature $T(r)$ of the main constituent substances of listed planets, moons, or asteroids, both as functions of radial distance r measured in astronomical units. The 'minimal mass' has been obtained from the observed mass by replacing the 'missing' hydrogen and helium under the assumption of Solar-System abundances. The mass distribution appears two-humped. All temperatures are consistent with the drawn-in law $T_2 = 1/r_{14}$

It is finally worth mentioning that besides the planets and moons, the Solar System contains comets and asteroids of icy, carbonaceous, stony, and iron-rich composition, as well as smaller bodies and dust made of the same materials. The mass distribution $\dot{N}(M)$ of their flow rates, determined both via in situ measurements by spacecraft and via crater statistics on the moon, is an approximate power law, all the way from molecules up to the asteroids, given by:

$$M^2 \dot{N}_M = 10^{-19\pm2} \, \mathrm{g/cm^2 s} \quad \text{for} \quad 10^{-18} \lesssim M/\mathrm{g} \lesssim 10^{18} \tag{1.6}$$

with $\dot{N} := dN/dt$, $\dot{N}_M := \partial\dot{N}/\partial M$.

Problems

1. Kepler's law for a circular orbit, $v^2 = GM/a$, relates the sum of the masses $M_1 + M_2 =: M$ to their separation a and their revolution period $P =: 2\pi/\Omega$, $\Omega = v/a$. Find the *revolution times* of two celestial bodies as functions of M and a. What is the minimum P for two balls of equal mass $M_\odot$ and mean mass density $\rho/\mathrm{g\,cm^{-3}} = \{1, 10^6, 10^{15}\}$, respectively – i.e. for {stars, white dwarfs, neutron stars} – whereby $a \geq R_1 + R_2$, $(R_j = \mathrm{radii})$?

2. What *radius* $R(M, \Delta t; T)$ needs a *star* of mass M in order to radiate a fraction $\epsilon = 0.1\%$ of its rest energy during the time interval Δt at the surface temperature T of the Sun? Of particular interest are solar values: $M = M_\odot$, $\Delta t = 10^{10}$yr.

1.3 The Milky Way

The Solar System is located near the midplane of the (cloudy) Disk of the Galaxy – or Milky Way, a *spiral galaxy* of Hubble type Sb – at a separation of ≈ 20 pc from the midplane, and at a separation of (8 ± 0.5) kpc from its rotation center [Reid, 1993; 1 parsec $= 3.26$ lightyears $= 10^{18.49}$cm]. Here it should be stressed that the Galactic Disk is not plane, rather warped, at angles of $\lesssim 20°$ on length scales $\lesssim$ kpc [Spicker and Feitzinger, 1986]. The local warp is well-known as *Gould's belt* – which contains the gas clouds and the young stars – though this interpretation is often traded for some unexplained explosion, of range 0.5 kpc. Its warping is the reason why the Milky Way forms a broad band in the night sky, rather than a narrow line. Even so, the midplane of the Disk is well defined locally by the (cold, heavy) molecular clouds, of average scale height some 50 pc (typical of the inner part of the Galaxy, rather than of the solar circle).

Masswise, the Milky Way consists at 90% of *stars*, of masses between 0.07 and 60 $M_\odot$, and luminosities mainly between $10^{-5}L_\odot$ und $10^2 L_\odot$, yet with record values up to $10^7 L_\odot$, see Fig. 1.2. Only 10% of the mass is presently in the gas phase. The gas is *warm* and shines at some 10^4K – the ionization temperature of hydrogen – whenever heated (and partially ionized) e.g. by nearby stars; if poorly heated, we observe the gas as *cold HI regions* of approximate temperature 10^2K – in particular in the light of the 21-cm radiation (of nuclear spinflip of H) – or as yet colder *molecular clouds*, of approximate temperature 10 K.

The Solar System revolves around the center of the Milky Way at a velocity of $v_\odot = 2.2 \times 10^2$km/s, corresponding to a revolution period of $P = 2\pi R/v_\odot = 10^{0.8+22.4-7.3}$s $= 10^{15.9}$s $= 10^{8.4}$yr and an enclosed Kepler *mass* of $M = r_\odot v_\odot^2/G = 10^{22.4+2\times7.35+7.2}$g $= 10^{44.3}$g $= 10^{11}M_\odot$. Here it should be stressed that the Kepler law is strictly valid only for spherically symmetric mass distributions; but the overestimate, resulting for an extreme disk-like mass distribution, amounts to less than a factor of 3. Stellar members of the Galaxy have been seen out to 60 kpc from its center; such a large (Galactic) volume may contain as much mass as $10^{12}M_\odot$. Our Milky Way thus belongs to the large galaxies in the Universe, for a mass distribution ranging between $10^7 M_\odot$ and $10^{14}M_\odot$ per galaxy.

Conspicuous inhabitants of the Galactic Disk, besides stars, are luminous gaseous clouds – so-called *nebulae* – which are irradiated by nearby, hot stars, or energized by their supersonic expansions: HII regions, planetary

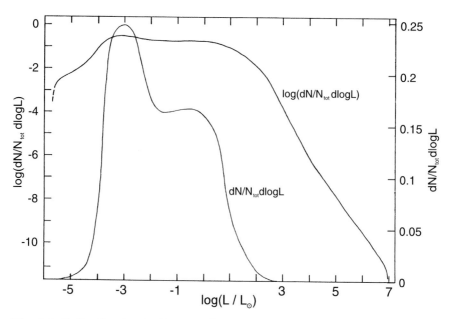

Fig. 1.2. Stellar Luminosity Function (of the local group of galaxies): plotted is the number of detected stars per logarithmic luminosity interval, $dN/N_{tot}d\log(L/L_\odot)$ vs $\log(L/L_\odot)$, both linearly (*right scale*), and logarithmically (*left scale*). Only a small percentage of stars has luminosities larger than our Sun; but such stars are shorter lived, hence have a larger time-integrated share in the population history of our Galaxy; see Fig. 8.2

nebulae,and supernova remnants (= SNRs). They differ from each other particularly in their spectra, both continuum and emission lines. There are also (massive) *dark clouds*, less irradiated, visible only at much lower frequencies.

Less spectacular are the (already mentioned) *cosmic rays*, an extremely relativistic plasma in apparent pressure equilibrium with the normal gas – or rather plasma – which apparently pervades the whole Milky Way, with ion energies starting slightly below rest energy and ranging all the way up to $10^{20.5}$eV, corresponding to Lorentz factors of $\gamma \lesssim 10^{11.5}$ for protons, a huge energy for an elementary particle, see Fig. 1.3. There is also a leptonic component of the cosmic rays consisting of both electrons and positrons, which involves perhaps a comparable amount of total energy, though distinctly less flux at high particle energies ($\gtrsim$ GeV, as opposed to particle Lorentz factors). When observers draw spectra in which the leptonic component is less energetic than the hadronic one, they assume a one-fluid structure of the Milky-Way plasma; such plots are not directly based on measured fluxes near Earth (for low electron energies), i.e. they depend on additional assumptions, and should therefore carry large (systematic) error bars. Moreover, the cosmic-ray transfer through the heliosphere ought to be charge dependent, because

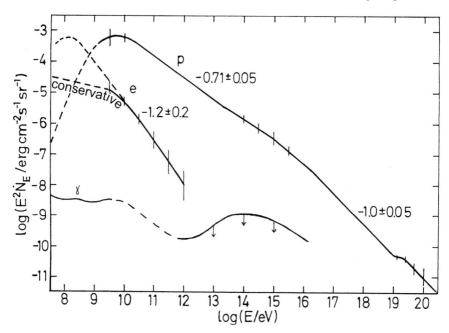

Fig. 1.3. Near-Earth Cosmic-Ray energy spectra, $log(E^2 \dot{N}_E)$ vs $logE$, for (dominantly) protons p, electrons e, and photons γ, with spectral indices and typical error bars indicated. At low particle energies, an attempt has been made to eliminate the 11-yr modulation by the solar wind. The 'conservative' electron branch assumes their homogeneous distribution throughout the Galaxy whereas the upper branch corresponds to the Milky Way being supported by pair plasma (whose bremsstrahlung is suppressed by being largely excluded from clouds)

of the higher rigidity pc/Ze of the (positive) ionic component; we lack direct information.

Among the inconspicuous inhabitants of the Galactic Disk range also the ubiquitous *magnetic fields*, of typical field strength 5 µG, hence of comparable pressure to the gas: $B^2/8\pi = 10^{-2\times5.3-1.4}\mathrm{dyn/cm^2} = 10^{-12}\mathrm{dyn/cm^2}$. They are about 50% ordered, almost parallel to the spiral arms.

What medium fills the Milky Way? This medium is commonly thought to be hydrogen, as the most abundant element in the Universe. But Reynolds [1990] finds a column density of warm hydrogen at most 20% of what would fill the available volume. Colder components ought to be denser (for supporting their ambient pressure) and could thus be observed, in (cold) emission or absorption. What about hot hydrogen, of temperatures between 10^5K and 10^7K? Arguments against contain the radiation temperatures which vary strongly from one line-of-sight to the next; apparently, such radiation comes from localised overpressure islands of small extent, like HII regions. Moreover, we shall find below that temperatures T between 10^4K and 10^8K are unstable, i.e. that matter with such temperatures either cools quickly, towards 10^4K, or continues heating up, towards $\gtrsim 10^8$K. The *hot* component which fills the Galactic Disk is probably relativistically hot: extremely relativistic *pair plasma*, of typical Lorentz factors γ up to 10^2 and more, jointly with part of the lower-energy ionic cosmic rays (which exert a comparable pressure, are expected to penetrate more deeply into clouds, and which are $10^{3.3}$-times smaller in number); see Fig. 1.4, and Plate 11.

Stars and clouds move relative to each other – in addition to their Galactic revolution – with *velocities* between ≈ 10 km/s and a few 10^2km/s, depending on mass, age, and type. The young, heavy stars, move most slowly, and the neutron stars most quickly (with $v \lesssim 10^{2.7}$km/s, a controversial number which has been recently raised by a factor of $\lesssim 4$, based on cases of uncertain distance, much to my worry). Escape velocity from the (potential well of the) Milky Way amounts to $10^{2.7}$km/s. The surface density σ of the

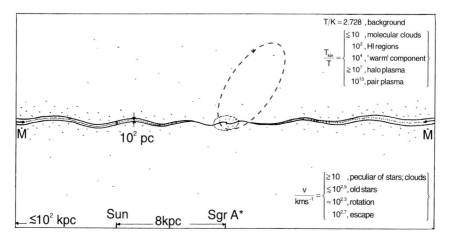

Fig. 1.4. Sketch of the inferred geometry of the Milky Way as seen from far away in the Galactic plane. Gas and dust of the Disk have a relative scale height of (only!) 1%, but warping makes the disk appear much thicker in projection. All population-I constituents, stars and clouds, oscillate through the disk, due to non-zero kinetic temperatures, whilst population-II objects (globular clusters, Schnelläufer stars) move along highly eccentric orbits with much larger amplitudes. Gaseous matter is inferred to spiral in towards the center, at a rate of $\dot{M} \lesssim M_\odot$/yr. Typical steady 'inhabitants' – beyond stars and clouds – are HII regions, supernova remnants, planetary nebulae, and synchrotron nebulae. The cosmic rays may escape from the disk through several 10^2 narrow 'chimneys'. Typical temperatures are 10^4K (of the 'warm' component), 10^2K (of HI), 10 K (of molecular clouds), $\gtrsim 10^7$K (of dilute halo plasma), and 10^{13}K (of pair plasma), all embedded in the 2.728 K cosmic background. All temperatures but the last are (only) kinetic temperatures. Typical velocities are between 10 km/s and $10^{2.5}$km/s of stellar and cloud peculiar motions, and $10^{2.3}$km/s of (ordered) Galactic rotation at not-too-small radial distances (of $\gtrsim 0.5$ kpc)

Disk's mass has been determined as $\gtrsim 10^{-2}\text{g}/\text{cm}^2$, corresponding to a typical gravitational field strength $g_\perp$ perpendicular to the Disk of

$$g_\perp = 2\pi G\sigma = 10^{-8}\text{cm}/\text{s}^2 \tag{1.7}$$

at the edge of the Disk; (for symmetry reasons, $g_\perp$ passes through zero in the middle of the Disk). Consequently, all objects oscillate perpendicular to the Disk, with a *scaleheight* H of

$$H \approx v_\perp^2/2g_\perp = 10^{14-0.3+8}\text{cm} \; v_7^2 = \text{kpc} \; v_7^2 \tag{1.8}$$

for vertical velocities in units of $10^2\text{km}/\text{s}$. For a gas of temperature T, the hydrostatic scaleheight in the same gravity field measures:

$$H = kT/mg_\perp = 10^{-15.9+6+23.8+8}\text{cm} \; T_6 = 3 \; \text{kpc} \; T_6 \;, \tag{1.9}$$

i.e. wants temperatures larger than 10^7K to fill the Galactic Halo. Gas below (a kinetic temperature of) 10^6K is gravitationally bound to the Disk, into a thin layer.

Velocities ought to be judged in relation to the respective *sound speed* c_s, i.e. the average speed at which molecules, atoms, or ions move locally. When the fluid medium obeys the adiabatic equation $p \sim \rho^\kappa$, with adiabatic index $\kappa = c_p/c_v = 1+2/f$ equal to the ratio of the specific heats at fixed pressure and volume (where $f :=$ number of degrees of freedom), c_s is known to obey:

$$c_s = \sqrt{dp/d\rho} = \sqrt{\kappa p/\rho} = \sqrt{\kappa kT/m} \overset{H}{=} 13 \; \text{km/s}\sqrt{T_4(m_p/m)} \;, \tag{1.10}$$

the last equality for hydrogen (H) at $T = 10^4\text{K}$, and with m as the mean particle mass. I like to memorise this result in the short form $c_8 \overset{H}{=} \sqrt{T_8}$; it will find repeated application.

For the warm component of the Milky Way ($T = 10^4\text{K}$), speeds of order 10 km/s are thus near sonic whereas for the cold component, sound speed is closer to 1km/s ; for (relativistic) pair plasma, on the other hand, we have $c_s = (2/3)c = 10^{10.3}\text{cm}/\text{s}$. *Supersonic speeds* are frequent in the Universe – where gravity can act through large distances – whereas they are rare on Earth, occurring (only) in: lightning, in a whip's crack, a supersonic aircraft, rockets, canons, guns, and bombs.

The composition of the Milky Way's *Halo* is not perfectly known. As calculated above, hydrogen would be a candidate if at $T \gtrsim 10^7\text{K}$; yet there is little evidence for such. For the space-filling agent of the Halo, better candidates are the pair plasma which escapes from the (magnetic fields of the) Disk, plus the ionic component of the cosmic rays. They escape on the timescale of $10^{7\pm0.3}\text{yr}$ – as measured by the lifetimes of their radioactive-decay products as well as by their secondary components – probably through Galactic *chimneys,* viz. small-scale leakages which are realized by escape channels rammed open via the overpressure of dense HII regions. (Remember that diffusive escape of air from an old air mattress can be tolerably slow, whereas such escape is prohibitively fast if due to a tiny puncture). In the escape process, some hot gas is likely to be dragged along, similar to sand storms which do not exclusively consist of warm air.

The *escaping number rate* $\dot{N}$ of relativistic ions should almost equal its generation rate, $\dot{N} = n\dot{V}$, with $n \gtrsim u/\langle\gamma\rangle mc^2 = 10^{-11.8-0.7+23.8-21}\mathrm{cm}^{-3} = 10^{-9.7}\mathrm{cm}^{-3}$ being the average number density of cosmic rays in the Disk, $u \approx \mathrm{eV/cm}^{-3}$ = their energy density, γ = their Lorentz factor, V = the occupied Disk volume (assumed box-shaped),

$$V = \pi R^2 H = 10^{0.5+45+20.5}\mathrm{cm}^3 = 10^{66}\mathrm{cm}^3 \ H_{20.5} , \tag{1.11}$$

and with the mean storage time $t = 10^7\mathrm{yr} = 10^{14.5}\mathrm{s}$ entering as $\dot{V} = V/t$, see Fig. 1.5. Consequently:

$$\dot{N} = n\dot{V} = 10^{-9.7+66-14.5}\mathrm{s}^{-1} = 10^{41.8}\mathrm{s}^{-1} \ H_{20.5} , \tag{1.12}$$

integrating to an ion number $N = 10^{59.3}H_{20.5}$ within $10^{10}\mathrm{yr}$, or more than $10^2 \mathrm{M}_\odot$ in mass.

This escaping cosmic-ray plasma suffices to *fill* the present *Halo* of the Milky Way. In order to see this, you could calculate the pressure which it exerts when distributed uniformly throughout the Halo. Even faster is the consideration that 10^{10-7} releases (within $10^{10}\mathrm{yr}$) from the box of height H would fill a truncated cylinder volume of height $10^3 H = 10^2\mathrm{kpc}$ at unreduced pressure, a pessimistic estimate.

Actually, a lot of neutral hydrogen is seen in the Halo, via its 21-cm radiation. Much of it is arranged as a band of falling *high-velocity clouds*, i.e. of cold clouds ($T \approx 10^2\mathrm{K}$) which rain down into the Milky Way, at free-fall speeds of $\lesssim 10^{2.3}\mathrm{km/s}$ which are distinctly smaller than if they came from outside of it (at escape speed, some $10^{2.7}\mathrm{km/s}$). Apparently, these clouds have formed via condensation in the Halo, after evaporation from the Disk, similar to dew on grass stalks. (The rising component is not detected.) Many of them are associated with tiny *intermediate-velocity clouds* which contain

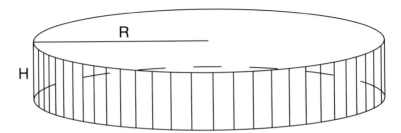

Fig. 1.5. Sketch of the simplified Galactic geometry to estimate the cosmic-ray storage volume of the Galactic Disk, assumed box-shaped

molecules; they could have resulted from the high-velocity clouds by having been blown at smoothly, with a light gas – the Galactic twin jet? – as their braking requires a large, smooth momentum transfer, at an inclined angle w.r.t. the Disk. This fountain-like phenomenon involves a mass rate of some $M_\odot$/yr, [Kundt, 1997].

Besides these (well-sampled) high-velocity clouds, the spectra also show a diffuse, spherically distributed component of high-velocity HI, of mass $10^{6.5\pm1}M_\odot$, mass rate $10^{-2\pm1}M_\odot$/yr, perhaps a true *fountain* phenomenon involving small filaments which have been shot out of the Disk by supernova explosions, and which subsequently fall back into it. Such filaments also make their appearance through the phenomenon of *refractive scintillations*, during routine surveys of distant point sources whose radiation can be systematically modulated (mainly reduced) during several weeks, most likely by a filament crossing the line-of-sight.

At this point, observers tell us the Solar System is sitting near the center of a *local hot bubble* (LHB), not exactly spherical in shape, of radius between 0.1 kpc and 0.2 kpc, which is essentially free of neutral hydrogen. Do we have to dismiss the Copernican principle? Should we not re-interpret the LHB as part of the multi-component Disk structure, with space filled by cosmic rays, both leptonic and hadronic, into which cloudlets of warm and cold hydrogen are embedded – like cirrus clouds in the troposphere, or (remotely) like tree stems seen from inside a wood – with 0.1 kpc as the typical distance between immersions? Galactic disks appear to be multi-structured, with cold condensates immersed in a hot matrix.

The Halo also contains stars, usually old ones ($\lesssim 10^{10.2}$yr), of *population II* – to be distinguished from the younger *population I* in the Disk which is more *metal-rich* – often concentrated in so-called *globular clusters*, i.e. spherical star clusters of mass $10^5 M_\odot$ to $10^6 M_\odot$. Their low spin suggests that they are the cores of formerly much more massive condensations ($\gtrsim 10^2$-fold). Their orbits do not lie in the plane of the Disk; they are highly eccentric ellipses, with the center of the Milky Way as their (near) focal point. Isolated Halo stars have similar orbits to the globular clusters, hence move faster than population I stars, and are known as 'Schnelläufer'; they may have been ejected during the formation of globular clusters. Population II is thought to have formed earlier than population I, from a more primordial gas, whence their low metallicities (of $< 0.1\%$, compared to $\gtrsim 2\%$ by mass); whereby in the astronomical nomenclature, all chemical elements beyond helium (of ordinal number > 2) are called *metals*.

The globular clusters are puzzling in various ways; there are indications that they are the debris of dwarf galaxies accreted by the young Milky Way. For instance: why do very few globulars contain large numbers of *ms pulsars* (13 from $10^{2.3}$, among them: 47 Tuc, M 15, M 5, M 13, Terzan 5, NGC 6624, ordered w.r.t. a falling number of pulsars) which pulsars should have turned off long ago, according to their age distribution in the Disk, and which should have been ejected at birth in the first place, due to a recoil which is thought to exceed escape velocity from the cluster core ($\lesssim (30-60)$km/s) [Kundt, 1998a]? The naively estimated excess ratio of neutron stars in a few globular clusters exceeds 10^6. Or why are there no planets around their stars?

Returning to stellar populations, there is a demand for yet another *population III*, viz. a massive, short-lived first population of stars which has provided (i.e. burnt and ejected) the metals necessary for populations II and (partially) I. Could they be identical with the vigorously burning, massive centers of *active* galaxies known as *QSOs* (= quasistellar objects), or *AGN* (= active galactic nuclei) ? The latter are known to eject strongly metal-enriched material, $\gtrsim 10^2$-times solar, the ashes of nuclear burning; they are commonly attributed to supermassive Black Holes (BHs) as the *central engines*, but are perhaps *burning disks* (BDs), the nuclear-burning cores of the galactic disks [Kundt, 2000]. Chapters 6 (on disks), and 11 (on bipolar flows) will deal with them.

What temperature can we assign to the Galactic Halo? In astrophysics, temperatures are almost exclusively determined from the spectra, by comparison with Planckian (or blackbody) radiation, see Fig. 1.6. A Planckian need not imply a uniform temperature in the emission region. But even if

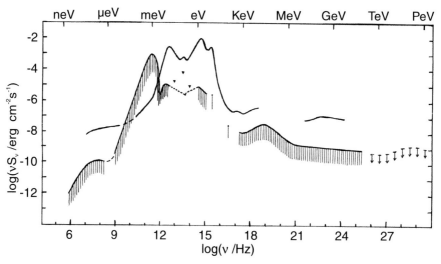

Fig. 1.6. Coarse spectra of both the Cosmic Background (per steradian; *vertical hatching*) and of our Galaxy, $\log(\nu S_\nu)$ vs $\log \nu$ or $\log E$, with large uncertainties wherever the foreground is bright and/or opaque. Note the energetic peak of the background near 1-mm wavelength, corresponding to 2.728 K, and the two Galactic peaks at visible and far-IR frequencies. Spectral luminosity L_ν and spectral flux S_ν of our Galaxy are related through $L_\nu = 10^{46}$ cm^2 S_ν. [see Nature 390, 257 (1997); Henry, 1999: Astrophys. J. 516, L49; Franceschini, Ausset, Cesarsky, Elbaz, Fadda, 2001: A & A. subm.]

the particles of a luminous medium have velocity distributions described by Maxwellians, i.e. have a uniform *kinetic temperature,* a true thermal equilibrium is rare: often the corresponding (large numbers of) photons are missing. More carefully, therefore, one should speak of *kinetic temperatures.* The true temperature of the Universe reads presently 2.728 K, corresponding to a microwave blackbody radiation which peaks at a wavelength of 1 mm, with $10^{2.6}$ photons per cm^3; it is omni-present, and contains almost the whole entropy of the cosmic substratum. Nevertheless, the Halo contains (sub-) populations of a multitude of (kinetic) temperatures, from molecular-cloud (10 K) through atomic (10^2K) and ionic ($\gtrsim 10^4$K), all the way up to extremely relativistic.

Problems

1. How long does pressure equilibration take through a distance d in a gas of temperature T? Calculate the *sound-crossing time* $t(d,T)$ for a) the solar wind ($d = 10^{13}$cm, $T = 10^{5.7}$K), b) the ISM ($d = 10^2$pc, $T = 10^4$K or 10^{13}K), and c) the IGM ($d =$ Mpc, $T = 10^7$K).

2. *Oscillation time* through the Galactic *Disk*: Calculate the free-fall time $P/4$ of a mass point from a height $\mid z \mid_{max}$ above a plane mass layer of thickness $2H$, homogeneous mass density $\rho = 10^{-23}$g cm^{-3}, with $H = 10^{2.5}$pc, for the cases a) $\mid z \mid_{max} \le H$, b) $\mid z \mid_{max} \gg H$; call $\sigma := 2H\rho$. Help: you have to solve the oscillator equation $\ddot{z} = -g_\perp(z)$ for the gravity acceleration $g_\perp$ obeying Laplace's equation $g'_\perp = 4\pi G\rho$.

3. For what particle energy $E = \gamma m_0 c^2$ (in eV) does *Larmor's gyration radius* $R_\perp = \gamma\beta_\perp m_0 c^2/eB$ of a proton in the Galactic magnetic field $B = 10^{-5.3}$G become comparable with the scale height $H = 10^{2.5}$pc of the Galactic Disk? Assume $\beta_\perp = \beta$; $\boldsymbol{\beta}_\perp := \boldsymbol{\beta} - \boldsymbol{B}(\boldsymbol{B} \cdot \boldsymbol{\beta})/B^2$, $\gamma := m/m_0$.

4. Are *galactic disks transparent* in the visible? A layer of optical depth $\tau = N\sigma$ transmits only $e^{-\tau}$ of all photons, where $\sigma = 10^{-24\pm1}$cm^2 is the mean absorption plus scattering cross section per particle (for the cloud-free, and nebula-free ISM at 10^4K), $N :=$ column density. Calculate $e^{-\tau}$ for a) lines-of-sight perpendicular to the Disk, for a mean particle density $n(z) = n(0)e^{-|z|/H}$, $n(0) = 10^{-1.5}$cm^{-3}, $H = 10^{2.5}$pc, b) roughly parallel to the Disk, though not through clouds, with $H = 10$ kpc, and c) through a cloud of thickness 10 pc, mean density $n = 10^5$cm^{-3}, with $\sigma = 10^{-23.5}$cm^2.

5. How does the gravitational lifetime Δt_g of a star (of mass M, contracting under its self-attraction) compare with its *hydrogen-burning lifetime* Δt_h (spent on the main-sequence)? For simplicity, assume (i) a constant luminosity $L = L_\odot(M/M_\odot)^{3.5}$, (ii) an available gravitational energy $\int pdv \approx GM^2/2R$, (iii) $R \sim M^{1/2}$, and (iv) a liberable nuclear energy of $10^{-3}Mc^2$, and express the result through the star's Schwarzschild length $\hat{M} := GM/c^2$.

6. Masses and densities of the bound *celestial bodies*: asteroids, planets, stars, white dwarfs, neutron stars, and black holes (if such exist) can be estimated from fundamental physics by equating the mean pressures which support and confine them. Approximate the repulsive *Fermi-Dirac pressures* and attractive {*electromagnetic/gravitational*} *pressures* by

$$p_F = \left\{ \begin{array}{ll} (3\pi/5)(\pi/3)^{1/3}\hbar^2 n^{5/3}/m \;, \text{NR} \\ (3/4)(\pi/3)^{2/3}\hbar\; c\; n^{4/3} \quad , \text{ER} \end{array} \right\}, \quad p \approx \left\{ \begin{array}{ll} e^2 n^{4/3} \qquad , \text{elm.} \\ G(M^2 m_p^4 n^4)^{1/3} \;, \text{grav.} \end{array} \right\}$$
$$\tag{1.13}$$

with $m_p :=$ proton mass, NR $:=$ non-relativistic. a) The typical laboratory density n_{elm} – equal at this approximation to the asteroidal density – follows from $p^e_{F,NR} = p_{elm}$, the typical planetary density n_{grav} from $p^e_{F,NR} = p_{grav}$, and the maximal density of {white dwarfs/neutron stars} from $p_{F,NR} = p_{F,ER}$ for {electrons/protons}. b) Chandrasekhar's maximal mass of cold stars follows from $p_{F,ER} = p_{grav}$, Fowler's maximal planetary mass from $n_{grav} = n_{elm}$, and the minimal black-hole mass from $M_{BH} \gtrsim nm_p R^3$ with $R \approx GM_{BH}/c^2$ (Laplace, Landau–Oppenheimer, Lynden–Bell). The results are suitably expressed by the nucleon mass m_p, number density n, fine-structure constants $\alpha := e^2/\hbar c$ and $\alpha_G := GM^2/\hbar c$, and by the Compton wavelengths $\lambda_j := h/m_j c$. c) An asteroid turns into a (spherical) planet when it yields to shear forces; the thus determined minimal planetary mass is smaller than Fowler's maximal one by over a factor of m_p/m_e.

1.4 World Substratum

What is our world made of? First in mass, and second in number – after the photons of the 3-K background radiation – ranges *hydrogen*, see Fig. 1.7. Modern cosmology often allows for non-baryonic, *exotic* matter as a possible abundant constituent, but so far, there is no direct evidence for this at all. Consequently, the world's substratum is far from its stable final state (at sub-nuclear pressure): *iron*. We live far from thermodynamic equilibrium. The chemical composition of Earth is far from typical.

We determine the composition of matter in the Universe from the spectra of stars, emission nebulae, clouds, as well as dispersed absorbers. *Helium*, the second most frequent element, is some ten times rarer in number than H, though only some 2.5 times rarer in mass. There was no chemical equilibrium in the dense initial state of the cosmos, soon after the *big bang*. In the model of the hot (i.e. thermalized) big bang – better called *big flash* – helium is formed from hydrogen during the first three minutes, in the rapidly expanding cosmic soup; it is formed much later in the model of the cold (non-thermalized) big bang, by population III stars. In the process of its formation, its huge binding energy of 7 MeV per proton is liberated; those sites should therefore belong to the brightest *cosmic sources*, brighter than present-day galaxies. Candidate

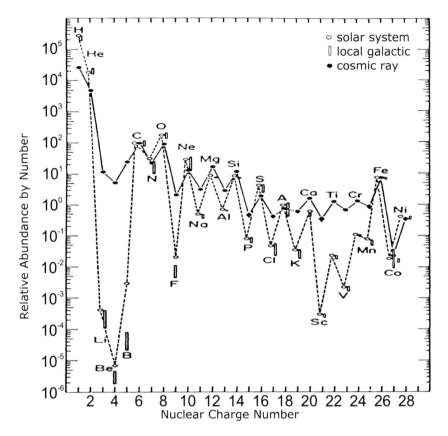

Fig. 1.7. Relative Abundances by number of the chemical elements, in the Solar System, in its local Galactic environment, and in the cosmic rays. Note that their average drop with nuclear charge number is exponential, that deep gaps in their distribution are filled up in the cosmic rays (by collisional spallations), and that H and He are depleted in the cosmic rays

sources are the 3-K background radiation, whose energy density must have been a larger – or even dominant – fraction of the cosmic substratum when it formed, because its energy density drops faster during adiabatic expansion than that of rest-mass-non-zero constituents; but so are the QSOs and quasars (quasi stellar radiosources), the nuclei of (radio-quiet and radio-loud) active galaxies.

The abundances of the heavy elements, beyond helium, roughly decrease exponentially with their ordinal number – with the exceptions of lithium, beryllium, and boron which are extremely under-abundant, by factors of order 10^{-7}. Moreover, there is an even-odd asymmetry w.r.t. ordinal numbers. All these abundances should be understood as the result of nuclear burning inside stars which subsequently eject (part of) the ashes of their burning in their winds, or explosively in novae, or in supernovae. This balance must not forget the QSOs, as the spectra of their BLRs (broad-line-regions) are spectra of nuclear ashes, exceeding solar metallicities (up to iron) by factors of $\gtrsim 10^2$. The BLRs are the ejection regions surrounding QSOs whose velocities reach large fractions (10^{-1}) of the speed of light.

In more detail, one distinguishes between *Solar-System abundances, local-Galactic abundances,* and the abundances of the *cosmic rays;* whereby Solar-System abundances can again be sorted into *solar surface* (or *chromospheric,* found from spectral analysis) and *solar wind* (collected by spacecraft): The more easily ionizable elements, with $\Phi_{ion} < 8$ eV, are some three-fold enhanced in the wind. Most prominent among the different distributions is the smoothness of the element distribution in the cosmic rays: Relative to carbon, hydrogen and helium are strongly depopulated (when compared with local Galactic), and all under-abundances (Li, Be, B, F, Sc,...) are filled up by the spallation products of neighbouring nuclei of higher ordinal number. Consequently, the cosmic rays have not simply been formed by random accelerations of the interstellar medium (ISM); selection processes must have held back the light elements (H, He), and those not easily ionized.

At this point, I am impressed by how friendly-for-life our planet *Earth* has been equipped (chemically). It is apparently not unimportant that the elements C, O, and Fe form relative maxima of the cosmic production.

1.5 Distance Ladder

Our knowledge about structures in the Universe depends crucially on our knowledge of their *distances,* because sizes, transverse velocities, densities scale with powers of distance and are often at the root of a correct interpretation. For instance, our 1999 estimates of the distances of the sources of the daily γ-ray bursts differ by factors of $\gtrsim 10^8$, from $\lesssim 0.1$ kpc to $\gtrsim 10$ Gpc, with corresponding differences in the involved energies by factors of 10^{16} (see Chap. 10). It is therefore important to use as many independent methods as possible for determining reliable distances.

At present, we know of *six* principally different *methods* to determine distances which together allow us to climb the ladder all the way up to the edge of the observable Universe. They are:

1. *Parallax method*: Stars which are near enough to project measurably onto different sky positions during a year, due to the Earth's revolution around the Sun, perform small yearly ellipses on the sky. Their distances d can be calculated from

$$d = v_\oplus / \mu \,, \tag{1.14}$$

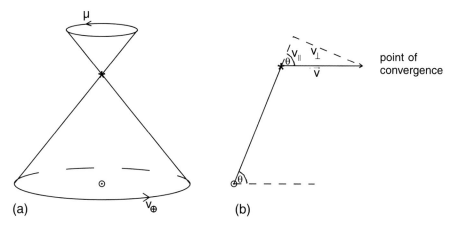

Fig. 1.8. (Apparent) Parallactic Motion of nearby objects in the sky: (**a**) stars, due to the yearly motion of Earth around the Sun, and (**b**) (kinetically cool, young) star clusters with a significant (uniform) bulk velocity relative to the solar system

where $v_\oplus$ is the velocity of the Earth around the Sun, and μ is the object's (2-d) angular velocity on the celestial sphere, or the observing telescope's angular velocity, called parallactic motion, see Fig. 1.8a. An overhead object has the distance of 1 pc when its yearly circle in the sky has a parallactic radius of one arc second: pc $=$ AU$/1'' = 10^{13.2+5.3}$cm $= 10^{18.5}$cm. Clearly, this method can only be applied to cosmic nearest-neighbours.

2. *Headlight method*: Whenever one deals with objects of known intrinsic luminosity L, one can use the headlight method familiar from judging the distance of approaching cars at night, on a straight road:

$$d = \sqrt{L/4\pi S} \,, \tag{1.15}$$

where S is the energy flux (= energy per area and time) arriving at Earth. This method has been applied to (i) *main-sequence stars*, burning H to He like our Sun, whose luminosity $L(T)$ is a known function of their observed temperature. It has been likewise applied to (ii) *Cepheïd* and *RR Lyrae* stars, viz. oscillating stars whose luminosity $L(P)$ has been found to be a unique function of their brightness-oscillation period P, (iii) the (ten) brightest stars of a galaxy, (iv) the brightest planetary (emission) nebulae, and (v) the brightest star clusters of a galaxy, all of which have similar luminosities for similar types of galaxies. Even (vi) the brightest galaxies in large clusters of galaxies have been used as *standard candles* for their distance determination. The best standard candles for cosmic distances may be (vii) *supernovae of type Ia* near maximum light, because of their huge luminosities and remarkable similarity, though even this small subclass of SNe does show non-uniformities in their spectra and lightcurves at various frequencies which signal individualities.

3. *Star-Stream Parallaxes*: Stars tend to be born in groups, or clusters, with more or less vanishing peculiar velocities (w.r.t. their center of mass). Due to the laws of perspective geometry, these stars appear to move in the sky along straight lines (great circles) all of which intersect in a distant point of convergence, at angular distance θ, see Fig. 1.8b. Their transverse velocity $v_\perp$, and line-of-sight velocity $v_\parallel$ obey $v_\perp = v_\parallel tg\theta$, so that their distance $d = v_\perp/\mu$ is given by

$$d = v_\parallel tg\theta/\mu \qquad (1.16)$$

as soon as $v_\parallel$ is known. But $v_\parallel$ can be measured via the Doppler effect whose general formula is provided by the Special Theory of Relativity as

$$\nu'/\nu = 1/\delta := \gamma(1 - \beta\cos\vartheta) \stackrel{\vartheta=0}{=} \sqrt{(1-\beta)/(1+\beta)} \approx 1 - \beta \qquad (1.17)$$

in which ν, ν' are the frequencies measured by the source and observer, respectively, whereby the source recedes at velocity $c\beta$, at an angle ϑ w.r.t. the line-of-sight, and where $\gamma := 1/\sqrt{1 - \beta^2}$ = Lorentz factor, $\delta :=$ Doppler factor. For small $\beta_\parallel := \beta\cos\vartheta$, this Doppler formula simplifies to

$$-\Delta\nu/\nu \approx \Delta\lambda/\lambda \approx \beta_\parallel \qquad (1.18)$$

with $\Delta\nu := \nu' - \nu$. The Doppler effect allows to determine $v_\parallel$ whenever the source emits, or absorbs at least one identified spectral line.

4. *Baade–Wesseling method*: When a luminous, expanding or revolving source has spherical or circular symmetry, a measurement of its maximal approach speed $v_\parallel$ implies a knowledge of its maximal transverse speed $v_\perp$, which is independently observed as a parallactic speed μ. With this, (1.14) becomes applicable; no nearer rungs of the distance ladder are required. This method has been applied to several resolved supernovae (for spherical symmetry), and recently also to masers on an inner Kepler orbit around the rotation center of the galaxy NGC 4258 (for circular symmetry).

5. *Cosmic expansion*: As noticed by Edwin Hubble already in 1929, distant cosmic objects recede from our Galaxy at speeds which grow, to first order, linearly with their distance. This general recession, or *Hubble flow* of galaxies is a consequence of the cosmic expansion, and is thought to be independent of the observer's position in space: at fixed cosmic time, all distances d grow roughly at the same rate $H(t) = \dot{d}/d$. The present value H_0 of the Hubble parameter H, $H_0 := H(t_0) = (70 \pm 20)$km/sMpc $= 10^{-17.65\pm0.15}$s^{-1}, is of order 1/age(Universe). Homo Sapiens owes his existence to this long time-scale, because biological evolution alone takes some $10^{0.6}$Gyr. Using (1.18), the distance d of a distant enough galaxy is approximately given by

$$d \approx cz/H_0 \quad \text{with} \quad z := \Delta\lambda/\lambda \approx \beta_\| \ . \tag{1.19}$$

Clearly, this method has an intrinsic uncertainty controlled by peculiar velocities $\delta\beta_\|$, which can be as large as 1%.

6. *Gravitational lenses*: Every local mass concentration can serve as a gravitational lens, by bending ambient light rays towards it, i.e. by making light signals detour – an effect which transcends Newton's and Maxwell's theory. In practice, 1 out of 500 distant sources is significantly enhanced by a foreground lens, by an arbitrary factor which tends to be several. If a lensed object is rapidly time-variable, its two or more images will vary with specific delays, independent of frequency, which can be read off their lightcurves. In this way, angular separations can be converted to linear separations, and their distances determined.

Problems

1. How sensitive must an optical spectrometer be in order to resolve line-of-sight velocities $c\beta\cos\vartheta$ of size a) 10 km/s, b) 10^2km/s? Express the *spectral resolution* through $\Delta\lambda$.

2. At what distance d does *Hubble's expansion velocity* $\dot{d} = H_0 d$ reach the magnitude of {stellar/galactic} peculiar velocities Δv (of $\{10^2/10^3\}$km/s)?

1.6 Galaxies, Clusters of Galaxies, IGM and Sponge Structure

In the sky, we do not only see our Solar System, in the foreground, and the Milky Way, in the middle ground, but also uncounted numbers of *galaxies* which arrange themselves more or less in clusters, or superclusters, or even in a large-scale ($\lesssim 10^2$Mpc) *sponge structure*, with underpopulated *voids* surrounded by sheets. The background of galaxies reaches to the edge of visibility, presently out to redshifts $z := \Delta\lambda/\lambda \approx 6$, corresponding to distances from where the wavelengths of emitted radiation arrive $(1 + z) = 7$-fold redshifted. At the time of such emission, the world was some $(1 + z)^{3/2} = 19$-times younger than it is now; see Plates 12 and 13.

A rule of thumb says that our Galaxy harbours 10^{11} stars, and the cosmos harbours 10^{11} galaxies. *Galaxy masses* reach down to $10^7 M_\odot$ – of the smallest detected dwarf galaxies – and up to $10^{14} M_\odot$, of the central cD-galaxies of large clusters which have significantly grown in mass – so we think – by galactic cannibalism. The *galaxy-luminosity function* is quite similar to the stellar one, Fig. 1.2; so far, the number of galaxies increases (slowly) with decreasing luminosity [MNRAS 2000: *312*, 557]. As a rule, we only see the (exponentially brighter) central part of a galaxy – comparable to the tip of an iceberg – and are often taken by surprise when at different wavelengths

(21cm, radio, X-rays), or in deeper maps, a galaxy reveals its $\lesssim 10$ times larger halo, and/or galactic associations.

Galaxies can have very different *morphologies*: The *Hubble sequence* arranges them linearly, as a tuning fork, with clean ellipsoids at the grip end, disks with spiral-arm structure with or without a *bar* in the two forked arms, and irregular galaxies at the other end(s); more careful classifications (by de Vaucouleurs, W.W. Morgan, S. van den Bergh) use 3-d, or multiply branched arrays. Spiral galaxies without a bar look like tropical storms (cyclones, hurricanes, typhoons) seen from outer space, see Plate 11.

Common to all galaxies is their composition, of gas and stars, the latter usually as constituents of a – more or less extended – disk. The gas (or rather plasma) in the disk rotates differentially, according to Kepler's (generalized) law, exerts magnetically mediated friction, thereby loses angular momentum to adjacent material further out, and spirals in towards the center, at average rates of $\lesssim M_\odot/\mathrm{yr}$. Consequently, galactic disks are not stationary – comparable to the (changing) cells of our body – but are replenished from outside, on the time-scale of cosmic evolution, often at different orientations. In galactic nuclei, matter therefore piles up repeatedly, resulting in (i) dense *molecular tori*, (ii) violent star formation (*starbursts*), (iii) coronal emission (*LINER*, = low ionisation nuclear emission region), and (iv) an *active nucleus* (*AGN*), all four of which add to the morphological appearance of their host galaxy, because of their non-ignorable brightness.

In addition, an active galactic nucleus can lead to the formation of two antipodal (*supersonic*) *jets*, in roughly 10% of all cases, more frequently so for high-luminosity sources. These jets ram vacuum channels into the circumnuclear and circumgalactic medium (CGM), of lengths up to several Mpc, of knotty appearance (with *hotspots*), and thereby blow *lobes*, or *cocoons* that belong to the largest, and strongest radio emitters in the Universe. Their cores tend to be observed as (compact) BLRs surrounded by (galaxy-scale) narrow-line regions (NLRs). Our highly controversial understanding of their functioning will be the subject of Chap. 11.

Galaxies often form groups, and associate in clusters, and in superclusters which reach sizes of 10 Mpc. The *superclusters* contain a lot of plasma which radiates at X-ray temperatures; one talks of *cooling flows* because the radiating plasma appears to fall into the supercluster from outside, thereby heating up in the outer zones (10^8K), and subsequently cooling in the denser core regions (10^7K). The inferred plasma densities n lie between $10^{-7}\mathrm{cm}^{-3}$ and $10^{-3}\mathrm{cm}^{-3}$. The peculiar velocities in large galaxy clusters can be quite high ($\lesssim 10^{3.3}\mathrm{km/s}$), and an application of the virial theorem to them – assuming they form bound systems – leads to masses which tend to distinctly exceed the sum of the visible masses of all components, by an order of magnitude; we deal with missing mass, with *dark matter*! Such mass estimates tend to be corroborated by estimates of their X-ray emitting cooling-flow gas, or by estimates based on their action as gravitational lenses, i.e. by their ability to focus the light of more distant sources. Is this dark matter hot (HDM), consisting of yet to be discovered (relativistic) elementary particles, or is it cold (CDM), consisting, perhaps, of clumped baryonic matter? The dark matter appears to concentrate in the cluster cores where the virial estimate yields an underestimate.

What medium fills the cosmic volume, the space between the galaxies? The approximate cigar shapes of the (lobes of the) extragalactic radio sources indicate that the stalled jet material meets with resistence as it expands subsonically into a medium of density $n \gtrsim 10^{-6} \mathrm{cm}^{-3}$, temperature $T \gtrsim 10^7 \mathrm{K}$, the latter from X-ray maps of large galaxy clusters. A lack of Lyα-edge absorption in the continuum spectra of background QSOs implies a much lower average neutral-hydrogen density, $n_H \lesssim 10^{-11.8}(1+z)^{3/2} \mathrm{cm}^{-3}$, hence again a high temperature of the volume-filling intergalactic plasma, $T \gtrsim 10^6 \mathrm{K}$, in order to have a high enough degree of ionization. I.e. the pressure of the *intergalactic medium* (IGM) is exerted by a rather thin and hot plasma, probably predominantly hydrogen. Note that a clumpy, cold component of the IGM would escape all our observations, for a wide range of grain sizes, hence cannot be excluded.

QSO spectra do, however, show hundreds of discrete Lyα absorbers, of column densities N_H power-law distributed according to $dN/d \ln N_H \sim N_H^{-0.5}$, between $10^{12} \lesssim N_H/\mathrm{cm}^{-2} \lesssim 10^{20.5}$, of inferred temperature $\approx 10^{4.3} \mathrm{K}$ and velocity dispersion $\gtrsim 5$ km/s – the *Lyα forest* – whose origin is ill understood. The extents of individual absorbers range from $N_H/n \gtrsim$ AU up to Mpc. Their metallicities Z are $\lesssim 10^{-3} Z_\odot$. But there are corresponding *metal-line absorbers*, occurring $\lesssim 30$ times less frequently, often hydrogen depleted, with a similar distribution. Both absorber systems are spatially distributed like the halos of glaxies, some tenfold expanded, with a similar joint chemical composition. They cannot be (statically confined) cloudlets because they would evaporate in subcosmic times. Marita Krause and I have interpreted them as filamentary supersonic ejecta from *active galactic nuclei*, diffusively segregated into metallic cores and hydrogen envelopes [1985: Astron. Astrophys. *142*, 150-156].

Out to what spatial scales does one detect inhomogeneities, i.e. structure in the Universe? Galaxy catalogues – including redshift measurements, for an estimate of their distances – have revealed a cosmic *sponge structure*, on the scale of $10^2 \mathrm{Mpc}$: The galaxies cluster with the geometry of a sponge, for redshifts $z \lesssim 6$, forming 2-d sheets which intersect in filamentary (1-d) edges, the latter perhaps meeting in 0-d vertices. In between the sheets are voids whose sizes would require speeds of order $10^2 \mathrm{Mpc}/10$ Gyr $= 0.03c$ if they were to be evacuated in the recent past, unrealistically large. This density structure must have been established in the early Universe.

Is there structure on yet larger scales (than $10^2 \mathrm{Mpc}$)? Among the split opinions discussed in the literature, the conservative one argues for asymptotic homogeneity [Wu et al, 1999] – so that the homogeneous-isotropic cosmological models of General Relativity become applicable – whereas a minority opinion argues in favour of a hierarchical, or fractal structure all the way up.

1.7 Cosmology

Our best theory of gravity is Einstein's theory of *General Relativity*. It generalizes Newton's theory to large velocities, and the Special Theory of Relativity to high mass concentrations; and it allows a straightforward embedding of Maxwell's theory of electromagnetism. Attempts have been made to even incorporate the electromagnetic, *weak*, and *strong* interactions between elementary particles into an unquantized, parameter-free $\geq$ eight-dimensional (Ricci-flat) generalization of it called *metron theory* [Hasselmann, 1998]. Einstein's field equations read:

$$G_{ab} = \kappa T_{ab} - \Lambda g_{ab} , \qquad (1.20)$$

with $\kappa := 8\pi G/c^4$ being the coupling constant between the 4-momentum properties of matter, described by the phenomenological stress-energy-momentum tensor T_{ab}, and Einstein's curvature tensor $G_{ab} := R_{ab} - (R/2)g_{ab}$, the latter composed of second space/time derivatives of the (dimensionless, 4-d) metric tensor g_{ab}; $R_{ab} :=$ Ricci tensor, $\Lambda := $ *cosmological constant*, an inverse area which acts like a (weak) repulsion and may, or may not vanish. Alternatively, cosmic repulsion could be the consequence of a non-vanishing electric charge density.

According to our knowledge, the Universe should be describable by a solution of (1.20) for all distances and times, including its beginning. The *conservation laws* of baryon number B, lepton number L, and charge Q, assumed to be valid throughout, allow us to apply them even to its much denser initial stage which may start with a singular beginning, the so-called *big bang* (after Fred Hoyle). For a non-negative non-gravitational energy density ρc^2 (with Λ incorporated into T_{ab}), such a big bang is unavoidable [see Kundt, 1972]. But a repulsive Λ, or a slight violation of the conservation laws (*continuous creation*) can avoid a singular beginning, in favour of an infinite sequence of bounded oscillations [Hoyle et al., 2000]. Unavoidably, all our conclusions turn uncertain with increasing distance from here and now.

Yet we are confident to eventually come within grips of our cosmic past. The mean mass density in the Universe should not be distinctly smaller than the *critical density*

$$\rho_{crit} = H^2/8\pi G = 10^{-29.4} \mathrm{g\,cm}^{-3}\, H_{-17.6}^2 \qquad (1.21)$$

which is obtained from the time-time component of (1.20) for a homogeneous-isotropic model with $\Lambda = 0$ (because of $G_{00} = H^2/c^2$, H as in (1.19)). This critical mass density corresponds to one hydrogen atom per m^3. We owe our existence to its smallness, as it has given Earth enough time for life to evolve.

The present value of the *density parameter* $\Omega := \rho/\rho_{crit}$ should not distinctly exceed unity (for the Universe to be old enough); it may be some 30-times smaller, even with dark matter included. With this, the world is $\gtrsim 10^{10}$yr old, has an observed extent (of the backward light cone) of $\gtrsim 3$ Gpc, and will probably expand for all times, i.e. not recollapse under its own attraction, (the hyperbolic case of expansion).

A basic fact of cosmology is the 2.728K background radiation, the cosmic carrier of entropy (or modern *ether*), whose 10^{-3} dipole anisotropy at the solar system tells us our state of motion w.r.t. the substratum. Its impressive *blackness* ($\gtrsim 10^{-4}$) and unexpected *isotropy* ($\lesssim 10^{-5}$) on all angular scales – with a peak of fluctuations at $1°$ – are embarrassing for all cosmological theories, already because of fluorescent re-emissions of Lyman-edge absorptions after decoupling (from matter). Has it been smoothed by a presently invisible scatterer, such as carbon or iron *whiskers*, or hydrogen snow before its vaporization?

What else can we learn from cosmology? We should like to know where and when hydrogen was burnt to the present-day distribution of the chemical elements, when the *background radiation* was formed and decoupled from matter, how the cosmic *sponge structure* came into existence, and how all the (various) *galaxies* formed: from above in mass (top down), or from below in mass (bottom up), or both at once? Unconventional answers to these problems are given by Hoyle et al. [2000]. Was the big bang *hot*, i.e. thermalized, or *cold*, i.e. initially photon-free [Layzer, 1990]? Do all the (generalized) *charges* initially vanish, like net charge Q, baryon number B, lepton number L, cosmological constant Λ? How did the *magnetic seed fields* form, which have led to equipartition strengths on galactic and even on cluster scales, and from which the stars, and the planets inherited their initial fields? Squeezed via shearing in galactic and protostellar disks, and subsequently blown out from their cores and stretched via winds, and bipolar flows? Note that ordered magnetic fields of strength $\gtrsim 10$ µG on scales up to $\lesssim 0.5$ Mpc (!) have been inferred from *rotation measures*

$$RM := (e^3/2\pi m_e^2 c^4) \int n_e \boldsymbol{B} \cdot d\boldsymbol{x} = \mathrm{cm}^{-2} \left(\int n_e \boldsymbol{B} \cdot d\boldsymbol{x} \right)_{16.7} , \qquad (1.22)$$

for the central regions of large galaxy clusters [Kronberg, 1994, 2000], via a frequency-dependent phase shift $\Delta\Phi = RM \, \lambda^2$ of linear polarization, see (3.14). The rotation measures can exceed 10^4rad/m^2 in the strongest *cooling-flow* clusters, whose infalling plasma glows at X-ray temperatures.

It is not clear how far we have gone yet in answering these cosmological questions reliably.

2. Gas Dynamics

Matter in the Universe is highly dispersed on average; yet we believe there is nowhere strict vacuum. The different components of gas and radiation try to fill the maximum available volume each, in obeyance of the second law of thermodynamics. They can thereby either penetrate each other, or – if immiscible – form separate domains in mutual pressure equilibrium. In quasi-static situations, pressure equilibrium is expected on length scales which can be traversed at sound speed since the last excitation: we speak of a *typical pressure* in the (outer) Milky Way, for example. This equilibrium pressure is permanently disturbed by the blowing of stellar winds – which generate reduced densities (near-vacua) at large radii – by the heating via stellar radiation (Strömgren spheres), by nova and supernova explosions as well as by bipolar flows. Supersonic motions lead to *shock waves* where they impact on slower substrata. The thus-created discontinuity surfaces can be smooth, when stable, or rough (rugged, filamentary), when unstable.

2.1 Galactic Pressures

A gas exerts a *pressure* onto the walls of its container by having its molecules elasticly reflected from them. The molecules transfer twice their normal momentum component in each collision. For a one-component gas, integration over all momenta and positions leads to the formula:

$$p/u = \begin{cases} 2/3, \text{ NR} \\ 1/3, \text{ ER} \end{cases} \qquad (2.1)$$

with $p = nkT$ = pressure, n = number density := number per volume, k = Boltzmann's constant, T = absolute temperature, u := energy density (per volume), and where {NR, ER} denote the non-relativistic, and extremely relativistic limit respectively.

For a gas consisting of several components, one has to sum over all *partial pressures* $p_j : p = \sum_j n_j k T_j$. Here the gas (volume) densities n_j are found, observationally, from their *column densities* $N_j := \int n_j ds$, either in absorption, or in emission, as well as from the *emission measures* $EM := \int n_e^2 ds$ of luminous *nebulae*. Such determinations of *average values* $\langle n_e^k \rangle := \int n_e^k ds$ $/ \int ds$ must allow for a clumpiness in their distribution, so that

$$\langle n_e \rangle^2 = f \langle n_e^2 \rangle \tag{2.2}$$

holds for a 2-valued distribution $n_e(\boldsymbol{x})$ which vanishes in subdomains, where the *filling factor* $f := V_{occ}/V \ (\leq 1)$ measures the relative occupied volume.

In the Milky Way, gases and plasmas coexist with the cosmic rays, magnetic fields, and electromagnetic radiation whose partial pressures add up to the total Galactic pressure. Magnetic forces are tensorial forces which pull parallel to the fieldlines, and push perpendicular to them; but in our rough estimate of pressures, we ignore their directional dependence, and put $p_B := B^2/8\pi$ as a representative for magnetic pressure. The resultant *Galactic pressure* then reads:

$$p_{gal} = p_{gas} + p_{CR} + p_{light} \pm p_B + p_{3K} \ . \tag{2.3}$$

Remarkably, all terms on the right-hand side (RHS) of this equation have comparable values: The pressure of the *warm component*, $p_{gas} = 2n_e kT = 10^{0.3-0.7-15.9+4} \mathrm{dyn/cm^2} = 10^{-12.3} \mathrm{dyn/cm^2} \ n_{-0.7} T_4$, results from a mean electron density $\langle n_e \rangle = 0.2 \ \mathrm{cm^{-3}}$ in clouds at $10^4 \mathrm{K}$, inferred e.g. from the *dispersion measures* $DM := \int n_e ds = N_e$ of pulsars, and also from emission measures. Cooler components have correspondingly higher densities which can grow almost unlimited, beyond equilibrium at zero gravity, after an onset of gravitational collapse.

The *cosmic rays* have an energy density of $\mathrm{eV/cm^3} = 10^{-11.8} \mathrm{erg/cm^3}$, whence a pressure $p_{CR} = u_{CR}/3 = 10^{-11.8-0.5} \mathrm{dyn/cm^2} = 10^{-12.3} \mathrm{dyn/cm^2}$, (2.1), which is comparable to that of the non-relativistic gas – by accident? I like to interpret this balance as a saturation: Cosmic rays are pumped into the Galactic disk like air is pumped into a leaky air mattress whose pressure saturates at some value which is controlled by the power of the pumps and by the amount of leakiness. The Galactic magnetic fields prevent an easy escape of the (gyrating) charges, and hence confine the cosmic rays in the disk except for the presence of a large number of small leakages, so-called chimneys, through which they can escape into the halo after a mean residence time of $10^7 \mathrm{yr}$; we will return to them in Chap. 10. Apparently, the Milky Way is pumped up by the cosmic rays – comparable to a cake's dough which is expanded, during baking, by evaporating baking powder – such that there is pressure balance between the driving and the residing gas.

Light pressure tends to be ignorably small in most terrestrial situations, but already the solar wind exerts less pressure than the Sun's light, by a factor of $\rho v^2 c/S_\odot = 10^{-3.1}$ (near Earth), quite noticeable by spacecraft, and by the cometary tails. From the measured spectrum of our Galaxy (Fig. 1.6), we find a typical starlight energy density $u = 4\pi S/c = 10^{-11.8} \mathrm{erg/cm^{-3}}$ in the Disk, whence a radiative pressure $p = u/3 = 10^{-12.3} \mathrm{dyn/cm^2}$, equal to the cosmic-ray pressure (within our rough estimate). Clearly, these two pressures can take very different values in different Galactic environments; but their comparability in the mean may tell us that similar energetics are involved in their production.

Galactic *magnetic fields* have been found to be comparable superpositions of an ordered component, roughly – though not strictly – parallel to the spiral arms, and a locally injected, chaotic component. The magnitude B of their vectorial sum scales as the galaxy's surface mass density σ, an expected property for pressure scaling as B^2 because of $p_\perp \lesssim \pi G \sigma^2$, see (2.4). For our Galaxy, a typical field strength of 5 µG corresponds to magnetic pressures $\pm p = B^2/8\pi = 10^{-10.6-1.4}\text{dyn/cm}^2 = 10^{-12.0}\text{dyn/cm}^2$, somewhat larger than the other pressures. This suggests that galactic magnetic fields have reached saturation, i.e. are steadily regenerated, mainly by being stretched by the shearing motion of the (highly conductive) orbiting plasma. Note that magnetic pressures can never exceed the yielding tensions of their dynamos: they would tear them, and behave as in vacuum. The field-anchoring clouds must exert pressures exceeding 10^{-12}dyn/cm^2.

There remains a discussion of the *background-radiation* pressure $p_{3K} = 10^{-12.9}$ dyn/cm^2 in the above formula: this pressure is uniform (within 10^{-5}) throughout the Universe – except inside Faraday cages which screen mm radiation – and is therefore dynamically uninteresting. Still, it dominates all (gas and light) pressures far from galaxies, as it is almost comparable to them inside their disks; it is the main carrier of cosmic entropy.

Moving substratum can exert *ram pressures* ρv^2 which exceed static pressures if and only if supersonic. Ram pressures are required, among others, for stretching magnetic fields; they are abundant in the Universe. They dominate inside windzones, HII regions, cloud cores, explosions, and will be discussed in the subsequent sections.

Does it make sense to ask for the pressure exerted by the *stars* in a galaxy, or of the dense *clouds*, treated as ensembles of point masses? Of course, it does make sense to ask for their energy densities, so why not pressures: dynamic friction couples them with the other components. Due to their huge masses, their kinetic temperatures $T = mv^2/3k$ are enormous. But their number densities are tiny, and the product of the two – believe it or not – is again similar to the above values. There is a simple reason for the approximate equality of their pressures:

$$p = nkT = \rho kT/m = \rho H g_\perp \tag{2.4}$$

holds, with $\rho :=$ mass density, $H := kT/mg_\perp =$ Galactic scale height, and $g_\perp :=$ gravity acceleration perpendicular to the disk. Because the stars and the galactic clouds move in the same gravitational potential with comparable scale height H as the dispersed gas, comparable pressures result for comparable mass densities. In our Galaxy, today's stars comprise 10-times more mass than the gas, hence have a 10-times higher pressure.

Problem

1. Calculate the *maximal magnetic field* strength B that can be anchored by a hydrogen plasma of a) static pressure $p = 2n_e kT$, b) ram pressure ρv^2, (due to $B^2/8\pi \leq p$). Of particular interest are the values $T = 10^4$K, $n \cdot$cm^3 $= \{10^{-1}$ (interstellar), 10^{18} (stellar photosphere)$\}$ as well as $v \lesssim 10^3$km/s (stellar winds, galaxies).

2.2 Shock Waves

Whenever a *supersonic motion* meets an obstacle, it drives a *shock wave*. Intuitively, the term 'wave' can be misleading: A shock wave is not an excitation propagating through a medium, it is a mass motion; the medium is discontinuously compressed, heated, and set in fast motion. It is swept away. Shock waves occur naturally as solutions of hyperbolic differential equations, even for continuous initial data, and hence differ from ordinary waves only in being (stronger and) discontinuous. For instance, a shock wave forms in a sufficiently long tube after a finite time when a piston is pushed into it at constant speed. Shock waves occur frequently in astrophysics: in stellar windzones, HII regions, SN explosions, and other violent interactions, on different scales and in different numbers, as will be discussed below.

In order to derive explicit expressions for the discontinuous jumps across a *shock front* in the hydrodynamic quantities of a one-component flow, approximated by an ideal gas, let us follow Landau and Lifshitz VI and describe the shock locally as a plane, 1-d flow, in the Galilean reference frame in which the discontinuity surface is presently at rest, see Fig. 2.1a. In this frame, a supersonic flow enters from the left, say, described by its velocity v_-, mass density ρ_-, pressure p_-, temperature T_-, and sound speed $c_- = \sqrt{\kappa k T_-/m}$. On transition through the front, these five variables change abruptly into corresponding ones, denoted by an index '+', which are determined by the three conservation laws of mass, momentum, and energy, i.e. by the three conserved expressions: $j := \rho v$, $p + \rho v^2$, and $w + v^2/2 = \kappa p/(\kappa-1)\rho + v^2/2$, where $w = $ enthalpy density, $\kappa := c_p/c_v$ is the adiabatic index, and where the adiabatic equation of state $p \sim \rho^\kappa$ has been assumed for the transition through the shock front. Outside that surface, on the other hand, the isothermal equation of state is used: $p_\pm = \rho_\pm k T_\pm/m$, so that the flows are described by three independent variables each. Writing $\mathcal{M} := v_-/c_-$ for the incoming Mach number, an elementary though lengthy calculation yields (*Rankine–Hugoniot*):

$$\frac{p_+}{p_-} = \frac{2\kappa\mathcal{M}^2 - (\kappa-1)}{\kappa+1} \rightarrow \begin{Bmatrix} 1 \\ \infty \end{Bmatrix} \text{ for } \mathcal{M} \rightarrow \begin{Bmatrix} 1 \\ \infty \end{Bmatrix}, \qquad (2.5)$$

and

$$\frac{v_-}{v_+} = \frac{\rho_+}{\rho_-} = \frac{\kappa+1}{\kappa-1+2/\mathcal{M}^2} \rightarrow \begin{Bmatrix} 1 \\ (\kappa+1)/(\kappa-1) \end{Bmatrix} \text{ for } \mathcal{M} \rightarrow \begin{Bmatrix} 1 \\ \infty \end{Bmatrix}, \qquad (2.6)$$

plus a corresponding relation for the temperature jump which follows from $T = pm/\rho k$.

It is instructive to consider the weak and strong limiting cases of these three jump relations, for the Mach number $\mathcal{M}$ going to $\{1, \infty\}$: For (2.5), we get $\{1, \infty\}$; i.e. no change at all for $\mathcal{M} = 1$, the trivial jump, yet an arbitrarily large jump in pressure for $\mathcal{M} \to \infty$. For (2.6), the corresponding limit reads $\{1, (\kappa+1)/(\kappa-1)\}$, with the explicit values $\{4, 7, \infty\}$, respectively for $\mathcal{M} \to \infty$ and $\kappa = \{5/3, 4/3, 1\}$. The latter three cases correspond to monoatomic gases in the NR and ER limit, and to an isothermal (i.e. fast cooling) shock. Note that the *compression ratio* achieved by a (non-cooling) shock is limited, often $\lesssim$ 4-fold, unless magnetic fields and/or relativistic particles are involved which allow for $\lesssim$ 7-fold compression; higher ratios would require solid-state devices.

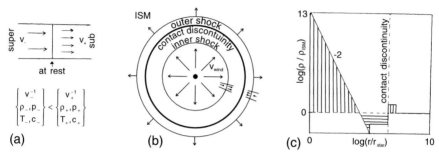

Fig. 2.1. (a) Shock-Wave geometry, **(b)** Stellar-Windzone geometry, and **(c)** radial run of mass density $\rho(r)$ of a stellar windzone, $\log(\rho)$ vs $\log(r)$. Shock waves are conveniently described in the rest frame of their discontinuity surface in which they satisfy the simultaneous inequalities $\{v_-^{-1}, \rho_-, p_-, T_-, c_-\} < \{v_+^{-1}, \rho_+, p_+, T_+, c_+\}$ for the supersonic pre-shock quantities (of index $-$) and corresponding subsonic post-shock quantities (of index $+$). In a windzone, the inner and outer shock surfaces have inverted orientations. Windzones are vacua in their outer parts; they sweep the formerly homogeneous ISM into the outer shocked shell. Wind matter and ISM touch each other at the contact discontinuity

In reality, a shock front is a thin 3-d layer with a finite *width*, thick enough for ample transfer of molecular momenta. Landau and Lifshitz derive from thermodynamic considerations a width of order a few mean-free-paths, or a few cyclotron radii in the presence of magnetic fields.

Problem

1. How fast – or at what Mach number $\mathcal{M}$ – must a gas of adiabatic index κ move in order to have its *pressure* {doubled/10^2-folded} when impacting on an obstacle? How large is the corresponding *density jump* ρ_+/ρ_-? Of particular interest is the case $\kappa = 5/3$.

2.3 Stellar Winds

Stars blow *winds* from their surfaces, by combinations of (i) hot upper atmospheres (coronae, with large scale heights), (ii) centrifugal forces, and (iii) radiation pressure. The relative importance of these three forces is not well understood, not even for our Sun; it depends sensitively on the radius out to which the escaping material is forced into *corotation* with the star, by magnetic rigidity; see [Lotova, 1988] for the Sun. This problem is extreme for the (hot, fast rotating) Wolf–Rayet stars, whose winds have radial momenta which can exceed those of their radiation by large factors: $\dot{M}c^2\beta/L =: \xi$ $\lesssim 30$, ($\beta := v_{wind}/c$). Such winds are often claimed to be driven by radiation pressure, for which each escaping photon would have to deliver $\gtrsim \xi$-times its outward momentum via multiple scatterings (on wind particles at opposite sides of the star). The difficulty may be compared to fleas in a match box whose jumps cannot make the box rise [see Kundt, 1998b].

Stellar winds can have quite different *strengths*: $10^{-14} \lesssim \dot{M}/M_\odot \text{yr}^{-1} \lesssim 10^{-4}$, and quite different velocities: $10 \lesssim v/\text{kms}^{-1} \lesssim 10^{3.5}$. Their ram pressure ρv^2 can sweep the ambient circumstellar medium (CSM) out to distances of $R \lesssim 30$ pc, depending on their strength, duration, and on the ambient density, and thereby create a cavity of reduced density, a quasi vacuum; its own *density* equals

$$n_{wind} = \dot{M}/4\pi r^2 vm = 10^5 \text{cm}^{-3}\ \dot{M}_{(-14)}r_{11}^{-2}v_{7.7}^{-1} \qquad (2.7)$$

with $\dot{M}_{(-14)} := \dot{M}/10^{-14}M_\odot \text{yr}^{-1}$ – being unity for our Sun – and with the often-used correspondence $10^{-8}M_\odot/\text{yr} = 10^{18}\text{g/s}$. At the distance of Earth from the Sun, $r = 1\text{AU}$, this formula yields a density $n_{wind} = 10^{0.7}\text{cm}^{-3}$, in agreement with the observed 7 protons/cm^3.

For a star at rest w.r.t. its CSM, its wind cavity will be more or less spherical, with its density and ram pressure dropping with r as r^{-2}. There will thus be a distance r_i where its supersonic radial motion is stalled, through an *inner (termination) shock*. Further out, at distance r_c, the stalled wind material meets the ambient medium at a *contact discontinuity* (stagnation surface) through which the composition can change (including its density) but not its pressure and speed. The surface of contact partakes in the radial expansion, and thus thrusts into the CSM; as a result, a second *outer shock* forms if this motion is supersonic, at a distance r_o where the unperturbed CSM first senses the build-up of the wind cavity.

In order to apply the results of the last section to the two spherical shock fronts just introduced, we have to switch to comoving frames: The wind certainly reaches the inner shock front at supersonic speed, and continues subsonically beyond, see Fig. 2.1b,c. In the rest frame of the outer shock (if it exists), on the other hand, the unperturbed CSM arrives supersonically from the outside, and is slowed down to meet the wind material at the contact surface; the (comoving) acceleration therefore happens in the opposite direction. In both cases, we may deal with highly supersonic flows so that strong shocks form, and the density jumps upward, by factors of $\lesssim 4$, towards the contact discontinuity. Note that the shocked wind material tends to be much hotter than the unshocked, or even shocked CSM, so that pressure equilibrium at the contact discontinuity can mean a (big) jump upward in density. Consequently, most of the accumulated mass lies in the shell outside of the contact surface; it is the compressed, *swept-up CSM* which formerly filled the cavity.

Most stars move considerably w.r.t. their CSM – i.e. feel an *interstellar wind* – so that their wind cavity will deviate from being spherical: subsonic motion distorts the sphere into an ellipsoid, whereas supersonic motion creates a *tail* in the downwind direction, like the Earth's magnetotail blown by the solar wind. We do not yet know which of these cases is realized by the *heliosphere* (blown by the solar wind): the Sun's motion, at $\approx 25\,\mathrm{km/s}$ w.r.t. the *local system of rest* (in the direction of 21 June, some 8° out of the plane of the ecliptic), is supersonic for warm hydrogen but would be quasi-static for (relativistic) pair plasma. At their speed of 3.5 AU/yr, the Pioneer and Voyager spacecraft may never reach the shock alive (at an expected distance of $\gtrsim 10^2\mathrm{AU}$). We observe neutral hydrogen and helium entering the heliosphere, being partially ionized and dragged along by the solar wind as *pickup ions*. Are they former members of a homogeneous interstellar hydrogen environment of the solar system, or are they immersed in a volume-filling pair plasma – like dust immersed in a dust storm – or in the form of sizeable filaments?

Problems

1. A star of mass M blows a wind of mass rate $\dot{M}$ at velocity v into a hydrogen plasma of temperature $T = 10^4\mathrm{K}$, pressure $p = 10^{-12}\mathrm{dyn/cm^2}$. What maximal *radius* r can its *windzone* reach? How long does that take? In particular, assume $M = 5\,\mathrm{M_\odot}$, $\dot{M} = 10^{-6}\mathrm{M_\odot/yr}$, $v = 10^3\mathrm{km/s}$. Help: for expansion, the wind's ram pressure ρv^2 must balance the ambient (static or ram) pressure.

2. What (geometric) thickness Δr has the (uncooled) *swept-up boundary layer* of the windzone from problem 1?

2.4 HII Regions

Hot, massive stars do not only push their environment (CSM) by blowing winds, but also by ionizing its hydrogen via their radiation (above the ionization energy of H, 13.6 eV), i.e. by blowing *Strömgren spheres.* Such Strömgren spheres tend to be much more extended than their contained wind-zones; they make their appearance as extended, luminous nebulae. Note that a star must be a lot hotter than the Sun for a Strömgren sphere to form. Extremely hot stars can even create HeII regions, requiring 24.6 eV, [Lang, 1998].

Radiative ionization and subsequent recombination involve strong heating, typically from some 10^2K to the ionization temperature of H, 10^4K; the pressure nkT thus rises by a factor of 200, as the number density n is doubled when H is decomposed into p+e. A Strömgren sphere thus evolves into an *overpressure* region – a bomb – which expands in the form of a strong spherical shock wave and tends to overtake the ionization zone at later times. The considerations of the preceding sections again find application.

Problem

1. The *Strömgren sphere* around a hot star expands at a velocity $\dot{r}_{ion}$ given by the absorption of its ionizing luminosity (power) L by the CSM: $L = 4\pi r^2 \dot{r} n E$, $E \gtrsim 13.6$ eV. At what radius r, and at what time t is it *outrun* by the overpressure wave launched by its ionization and heating, of pressure $p = \xi n k T$, speed $c_s = \sqrt{dp/d\rho}$? Assume $L = 10^{37}$erg/s, $n = 10^2$cm^{-3}, $\xi T = 10^{4.3}$K. (This simple estimate ignores the wind which simultaneously leaves the star).

2.5 Stability of a Contact Discontinuity

The boundary between two fluid media can be stable, like the surface of a lake, or unstable, like the inverted situation: Try to pour a pail-full of water from a window in the $\geq 3^{rd}$ floor onto somebody's head; you will be disappointed! In the unstable case, a *contact discontinuity* spreads into a 3-d layer of interpenetration of the two formerly adjacent media.

This situation can be generalised, from static gravity (in the case of the lake) to any accelerated layered medium, by replacing the acceleration of gravity, $\boldsymbol{g}$, by an arbitrary *acceleration* $\boldsymbol{a}$. The generalization reads: a boundary layer between two media of different mass density ρ is *stable* if and only

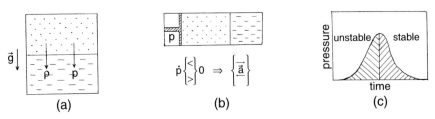

(a)   (b)   (c)

Fig. 2.2. Rayleigh–Taylor instability: (**a**) water basin in air, (**b**) steadily moving piston in a tube, and (**c**) inferred stability epoch, viz. the switch-off phase of a fluid medium compressing a heavier one, *hatched upward*. Stability during acceleration $\boldsymbol{a}$ wants the gradients of p and ρ to be parallel

if the gradient of pressure has the same direction as the gradient of inertia: $\nabla \rho \parallel \nabla p$, see Fig. 2.2.

In application to an explosion, the gradient of pressure points inward immediately after ignition, and reverses its direction later, when a quasi-vacuum forms at the center. We thus learn that when a *light medium* pushes a heavier one, the boundary layer between the two is (*Rayleigh–Taylor*) *unstable during switch-on*, and stable thereafter, during relaxation. This insight will have various applications, to (super-) novae, jets, etc.

2.6 Pressure Bombs

Bombs come in two varieties: *pressure* bombs, and *splinter* (*shrapnel*) bombs. The former are thin-walled, hence transfer most of their energy to the ambient medium, whereas the latter are thick-walled, hence transfer most of their energy to the fragments of their former case. Clearly, the latter have a larger range, with the compromise of having a patchy impact, leaving a lot of structure undestroyed; see Fig. 2.3. Hydrogen bombs in air are the best example of pressure bombs whereas supernovae are at the other extreme, a fact which has been overlooked by Shklovskii in 1962, and by most subsequent textbooks and research work.

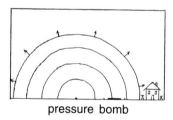
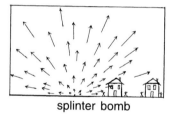

pressure bomb   splinter bomb

Fig. 2.3. Thin-walled bombs are pressure bombs, thick-walled bombs are splinter bombs; they cause qualitatively different damages

A pressure bomb can be approximated by an expanding spherical air mass whose kinetic energy $E = M(r)v^2/2$ is initially conserved. Putting $M(r) = (4\pi/3)\rho r^3$ – and trying the similarity ansatz $v = r/t$ – one finds for the shock radius r as a function of time:

$$r = (Et^2/2\rho)^{1/5} \sim t^{2/5} , \tag{2.8}$$

the well-known *Sedov–Taylor* wave which has been confirmed during nuclear-bomb tests. In particular, the shock slows down in proportion to $v \sim t^{-3/5}$.

A Sedov–Taylor wave does not conserve radial momentum which scales as $M(r)v$; it violates collision dynamics as ambient mass is swept up, by ignoring *thermal losses*. Such thermal losses are compensated, for some time, by the pressure of the enclosed photon bath which stabilises the expansion, but must be taken care of once the expanding air shell becomes transparent. Thereafter, v drops faster than $\sim t^{-3/5}$.

In the literature, Sedov–Taylor waves are routinely applied to *supernova remnants*, with dubious success because (core-collapse) supernovae are extremely thick-walled, and hence behave like splinter bombs. Each (gaseous) splinter, or filament expands as it slows down, due to a decreasing ram pressure, which leads to accelerated (exponential) slowdown.

2.7 Supernovae

When a massive star has burned the hydrogen in its core to helium, the core contracts, heats up, and ignites further nuclear reactions, with higher energy thresholds. In the *Hertzsprung–Russell diagram* ($logL$ vs $log(T^{-1})$), the star then leaves the main sequence, and climbs up the *giant branch*. Sooner or later, the *core* evolves towards pure iron – the tightest bound chemical element – cools, and requires the Fermi pressure of its degenerate electrons for its support against gravity: it becomes a *white dwarf*.

What happens next depends on the amount of mass that has remained on the star. Low-mass stars never make it beyond helium and are thought to end as white dwarfs: degenerate, stable celestial bodies which are born hot and cool to become black dwarfs. Stability against gravity requires their mass to be less than *Chandrasekhar*'s limit

$$M_{Ch} = m_p \alpha_G^{-3/2} = 1.4 \ M_\odot \ , \tag{2.9}$$

with $m_p :=$ proton mass, and with the gravitational fine-structure constant $\alpha_G := Gm_p^2/\hbar c = 10^{-38.229}$; see problem 6 of Sect. 1.3. The precise value for M_{Ch} depends on the chemical constitution of the cold star and can lie $\gtrsim 10\%$ below 1.4 $M_\odot$.

Once the star gets heavier, e.g. by accreting mass from a nearby companion (and not blowing it off again), it is doomed to collapse to become a *neutron star*, i.e. a star which is (thought to be) mainly composed of free neutrons. The degeneracy pressure of the free neutrons is deemed strong enough to stabilize a *neutron star* of mass $\lesssim 3 \ M_\odot$, whereby the exact upper mass limit depends sensitively on the state of matter near nuclear densities. Note that according to the non-linear equations of General Relativity, pressure has weight and eventually cannot prevent a collapse, no matter how strong it is. The result of such a collapsing superheavy star is thought to be a (stellar-mass) *black hole* which may, however, form extremely rarely because of natural hurdles, among them the Eddington hurdle, (6.12). We thus arrive at the scheme:

$$M \begin{Bmatrix} < \\ > \end{Bmatrix} M_{crit} \longrightarrow \begin{Bmatrix} white\ dwarf \\ neutron\ star\ or\ BH\ or\ \emptyset \end{Bmatrix} \quad via \quad \begin{Bmatrix} PN \\ SN \end{Bmatrix} , \quad (2.10)$$

whereby a *planetary nebula* (PN), or *supernova* (SN) results because of the liberation of the huge binding energy of the collapsing core. As the gravitational binding energy scales as R^{-1}, it is some 10^3-times larger for a neutron star than for a white dwarf. In (2.10), the entry $\emptyset$ stands for no compact remnant at all, i.e. for a total disruption; which should be a rare event because the *birthrate* of neutron stars looks (at least) as large as the SN rate, one in ten years in the Galaxy (when account is taken of the fact that all the historical supernovae have occurred within a $\lesssim 10\%$ vicinity of the Solar System).

This birthrate estimate is by no means commonplace. For radio pulsars it follows from their average lifetime $t = 10^{6.4}$yr, Galactic number $N \gtrsim 10^{5.1}$ (after correction for incompleteness), and beaming fraction $f \leq 1$ [Kundt, 1998a] as

$$\Delta t = t\ f\ /\ N \overset{PSR}{\lesssim} 20\ f\ \text{yr} . \quad (2.11)$$

Statistically, for every radio pulsar there is an older neutron-star brother, often observed as a binary X-ray source but sometimes invisible (when ejecting), yielding a neutron-star *birth interval* of $\Delta t \lesssim 10\ f$ yr. Note that the existence of an older brother is independently inferred from the large peculiar velocities of pulsars, partially inherited via binary fission.

The critical mass M_{crit} in (2.10) which determines whether a star eventually turns into a white dwarf or something more compact – if such a mass is well defined – is equally controversial. It should be consistent with (i) the birthrate of white dwarfs, (ii) the birthrate of neutron stars, (iii) the PN rate, (iv) the SN rate, (v) the supernova remnant (SNR) rate, and (vi) the *initial mass function* (IMF) which counts the number of stars as a function of their mass at birth. In view of the many neutron stars in the Galaxy – detected as pulsars, binary X-ray sources, or even invisible (when screened, without accreting) – I favour a *critical mass* of some 5 $M_\odot$ (over larger values, like 8 $M_\odot$). The bias would become even more severe if a large number of massive stars would end up as black holes (BHs); in my own judgement, none of the BH candidates (BHCs) do involve BHs, rather they are neutron stars surrounded by massive disks [Kundt, 1998a,b]: The proposed BHCs have too much spectral and variability structure, reminiscent of a rotating inclined magnet at their center, see Sect. 9.2.

For the rest of this section, let us concentrate on *core-collapse supernovae*. Other forms of SNe are likewise considered in the literature – like explosive disruptions of white dwarfs for type Ia – but need not be realistic: All the different SN types, classified by the different chemistries revealed by their spectra as of type {Ia, Ib, Ic, IIP, IIL, IIb, IIn}, show a great many similarities and transitions, reminiscent of just one explosion mechanism blowing off multi-layered envelopes of various mass, extent, and chemical composition.

In particular, the less-extended envelopes of blue supergiants require more ejection energy than those of the red supergiants, because of a larger escape energy, hence lead to similar but fainter lightcurves. Perhaps, the different *types* of SN are fully determined by the size and chemistry of their progenitors' envelope, coarsely as:

$$
\begin{array}{ccc}
 & red & blue \\
He & Ia & Ib \\
H & IIP & IIL
\end{array}
\qquad (2.12)
$$

May I warn the reader that mainstream treatments of SN explosions deviate from the interpretation in this section, see Burrows [2000] who admits, however, that "revitalising" an "accretion shock" is "a riddle wrapped in a mystery inside an enigma".

How much energy is available when a neutron star forms? The *gravitational energy* of a neutron star, of mass some 1.4 $M_\odot$, should be calculated within General Relativity because of the star's high compactness but does not come out far from its Newtonian value: $E_{grav}(n*) \approx 0.2Mc^2 = 10^{53.8}$erg $(M/1.4M_\odot)$, see Landau and Lifshitz V. Part of this energy is required to compress the star's matter against its elastic repulsion, towards nuclear densities. Their difference, the star's *binding energy*, results as $E_{bind}(n*) = 10^{53\pm0.5}$erg; it will be liberated during core collapse, in the form of heat that can drive nuclear reactions, and is thought to escape mainly in the form of neutrinos which are emitted during neutronization.

Another form of energy that will be liberated during a core's collapse is its *rotational energy* $E_{rot}(n*) \approx I\Omega^2/2 = 10^{52.7}$erg $I_{45}\Omega_4^2$, where a minimal spin period $P_{min} \lesssim$ ms has been assumed, consistent with the pulsar-spin record of 1.56 ms. A significant fraction of this rotational energy can be tapped via magnetic coupling of the accelerating (collapsing) core to the overlying envelope. The field will thereby be wound up toroidally, towards strengths of order 10^{16}G, and transfer most of the core's (extreme) angular momentum, on the time-scale of seconds. The field will reconnect soon thereafter and decay into a relativistic cavity – mainly electron–positron pair plasma – which can serve as the *piston* for transferring the liberated energy to the extended star's envelope and for building up its outward momentum. (Part of the piston may, but need not, reach escape velocity).

Supernova shells tend to have masses ΔM of order 3 $M_\odot$ – inferred from the times at which their spectra change from optically thick (*photospheric*) to optically thin (*nebular*), usually between 6 and 18 weeks after launch – and radial velocities ranging from hundreds of km/s up to several 10^4km/s, with a quadratic mean near $10^{3.8}$km/s. Their *kinetic energies* are thus of order $E_{kin} = \Delta Mv^2/2 = 10^{51}$erg, some 1% of the liberated energy. Yet much smaller are their *radiated energies* $\int Ldt = 10^{49.5\pm0.5}$erg, a few percent of their kinetic energies. Apparently, supernova shells deposit most of their explosion energy in heating and expanding their environment – a galactic disk

– whence it is retrieved during cloud contraction and eventually radiated in the infrared part of the spectrum.

The difficulty in numerically modelling a SN explosion lies in correctly modelling the transfer of the energy – liberated on the length scale of 10^6cm – to the envelope, of radius $10^{13\pm0.5}$cm, i.e. through a factor of 10^7 in radius, or 10^{21} in volume: SNe are extremely thick-walled bombs. The (gaseous) *piston* whose pressure transfers the *energy and radial momentum* cannot be non-relativistic, because it would cool by adiabatic expansion, on its runway out to the envelope, in proportion to $\rho^{\kappa-1} \sim r^{-2}$ (for $\kappa = 5/3$), by a factor of 10^{-14}. Its sound speed would drop to zero, and with it its capability to keep pressure contact with the envelope that recedes at $\lesssim 10^4$km/s. In contrast, a relativistic piston cools only as r^{-1} (for $\kappa = 4/3$), by a factor of 10^{-7}, hence retains a relativistic speed if starting at high enough Lorentz factors, $\gamma_{initial} \gtrsim 10^8$, Fig. 2.4.

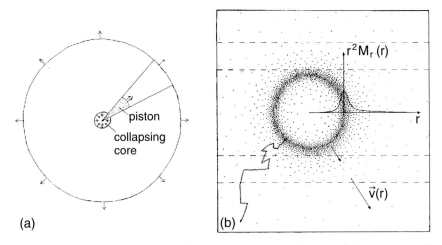

Fig. 2.4. Supernova explosion: (**a**) some relativistic medium, called piston, expels the envelope of the progenitor star during and after core collapse and decomposes it into a huge number of splinters (shrapnel, which perform a Hubble flow), because of a Rayleigh–Taylor instability. (**b**) The smoothed mass distribution in the shell of ejecta has a double-power-law bell shape, $rM_r \sim \{r, r^{-5}\}$, as inferred from the early spectra and from iterated maps during the remnant stage. The *broken lines* delineate zones which appear optically {thick, thin, thin even in lines} to a distant observer in their direction. Line photons, being multiply scattered, can be trapped for several months

What does the SN piston consist of? Most of a SN's energy is thought to escape as *neutrinos*, for which the stellar envelope is transparent; neutrinos thus do not qualify as a piston. Next, there is the hot *photon* bath; but photons mix with matter, and their mixture has a sub-relativistic sound speed (for the available energies). The only other available relativistic media I can think of are *magnetic fields* and their decay product: a *relativistic cavity*. In any case, a low-inertia piston implies – according to our above stability considerations – that the transfer of the 4-momentum should be strongly Rayleigh–Taylor unstable, and lead to a decomposition of the envelope into huge numbers of splinters, probably *magnetised filaments*. A SN is not a pressure bomb, it is a splinter bomb [Kundt, 1990a].

What is the expected radiation from a SN shell, plotted as *SN lightcurves* at various frequencies, Fig. 2.5? It should take a few hours until the piston reaches the outer edge of the envelope (where the latter becomes transmittent). During its sudden acceleration to terminal speed, the shocked matter should reach kinetic temperatures of $10^{9.3}$K $v_9^2 (m/m_p)$, according to (2.6), which imply Planckian temperatures of $\lesssim 10^{6.5}$K on energy sharing with the photon bath. A SN should therefore start with a UV flash of this temperature, lasting a shock-crossing time of the photosphere, about one ms, at the (huge) luminosity

$$L = 4\pi r^2 \sigma_{SB} T^4 = 10^{47} \text{erg/s } R_{13}^2 T_6^4 \; . \tag{2.13}$$

Such a short UV flash can serve like a flash light in photography and has been observed in the form of the *light echo* from SN 1987A in the LMC, (in the shape of Napoleon's hat). Its time-integrated energy is small: $\int_{flash} L dt \ll \int_0^\infty L dt$, but it can tell us a lot about the SN's CSM.

During the weeks and months after a SN explosion, its light $L(t)$ tends to rise because of a growing photospheric area $\sim R^2(t)$, but to decline because of a falling temperature, $\sim T^4(t)$. As a result, a SN may be detected early enough for $L(t)$ to first fall and soon rise again, towards its – usually well-sampled, second bolometric – *maximum*, whereupon the luminosity mostly falls in the shape of two exponentials, a steep one followed by a shallower one. Mild exceptions are SNe of type IIP. These e-folding decay times are of order 10 d and 10^2d, respectively, reminiscent of the radioactive decay of ^{56}Ni to ^{56}Fe via ^{56}Co, but can deviate therefrom by factors of 2 and more; radioactive heating is certainly involved, but mostly fails quantitatively, by factors of several. After about two years, a SN tends to be lost from sight, though in nearby cases it is often monitored for another $\gtrsim 10^4$yr, at a more-or-less constant luminosity of $\lesssim 10^{38}$erg/s, as a SNR.

What powers the *lightcurve* of a SN? Clearly, a SN starts as a glowing, cooling ball of ejected hot plasma, with an enclosed, cooling hot-photon bath whose temperature falls within days, from some 10^6K down to $10^{3.8}$K but hardly any further; thereafter, the temperature of the photosphere stays frozen, near the recombination temperature of hydrogen. The colour temperature even remains that high beyond the last break in the exponential dimming law, where the spectrum changes from *photospheric* to *nebular*: Diffusive line-photon leakage can simulate an exponentially decreasing thermal emission,

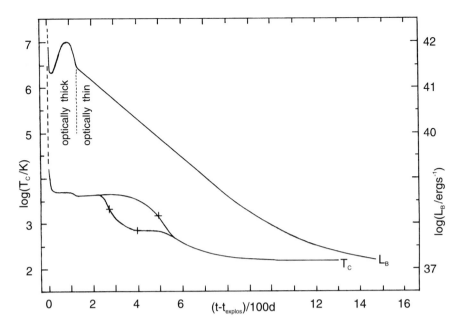

Fig. 2.5. The Bolometric Lightcurve $L_b(t)$ of SN 1987A and its colour temperature $T_c(t)$ during the first four years after launch. Note that T_c differs from the inferred (multiple) kinetic temperatures at late stages

near $10^{3.8}$K, whilst dust formation can be read off the spectrum, signalling much colder temperatures somewhere in the interior.

But the integrated radiation power exceeds the initial thermal energy of the exploded star by an order of magnitude; what extra energies are converted to light? As mentioned before, the literature stresses radioactive injection, from ^{56}Ni. There are, however, at least three further potential *energy reservoirs* in a SN explosion: (i) overtaking crashes between ejected filaments, when inner ones initially escape at higher speeds than outer ones; i.e. a tapping of the huge kinetic reservoir; (ii) the cooling pair-plasma piston which has launched the SN, and (iii) the non-neutrino cooling energy from the ($\gtrsim 10^8$K hot) central neutron star, which is predicted to be minor for conductive cooling of the star but could be important for convective cooling, via volcanoes.

Problem

1. At what time (after ignition) $t = r/v = t(\Delta M, \sigma)$ and at what radius r does a *SN shell* of mass ΔM ($= 3$ M$_\odot$) become *transparent* in the visible if the mean scattering cross section σ per proton takes the value 10^{-24}cm^2? The velocity v should be found from $\Delta M v^2/2 = 10^{51}$erg.

3. Radiation and Spectra

There would be no knowledge in astrophysics without the reception of radiation: electromagnetic, neutrino, gravitational, and cosmic rays, the latter forming an exception in having (definitely) non-zero rest mass. This chapter attempts to collect textbook results on electromagnetic radiation which are required for an analysis of the observed intensities and spectra. Derivations will start in a rigorous manner but proceed in a sketchy fashion, with the goal of providing a broad survey.

3.1 Radiation by an Accelerated Point Charge

An electron lying on our table must not radiate even though it is accelerated, by the Earth's attraction. It must not radiate because it would lose mass, whereas we know that all electrons have the same mass. We also know from Einstein that a freely falling electron (by implication, far from other charges) does not radiate, because it realizes the ground state of kinematics: the unaccelerated state, represented by a geodesic in 4-d spacetime. This seeming paradox – of a radiation-free electron – is not real: a *uniformly accelerated charge* does *not radiate*.

The last statement is an exact consequence of the (*Abraham–Laue*) *Lorentz–Dirac* equation of motion of a point charge, given by

$$\overset{\circ}{u}{}^a = (e/m_0 c) F^{ab} u_b + r^a \ , \tag{3.1}$$

in which $u^a := \gamma(1, \boldsymbol{\beta})$ is the (dimensionless!) 4-velocity of the charge, $\overset{\circ}{u}{}^a := du^a/d\tau := \gamma \dot{u}^a := \gamma du^a/dt$ its derivative w.r.t. proper time τ, or proper acceleration, m_0 its rest mass, $F^{ab} (= -F^{ba}) = $ the (antimetric) electromagnetic field tensor – having the electric-field components as space-time components, and the magnetic ones as space-space components – and in which the radiative reaction term

$$r^a := \tilde{\tau}(\overset{\circ\circ}{u}{}^a - u^a \overset{\circ}{u}{}^b \overset{\circ}{u}_b) \ , \quad \tilde{\tau} := 2e^2/3m_0 c^3 \overset{e}{=} 10^{-23.3} \text{s} \ (m_e/m_0) \tag{3.2}$$

is perpendicular to u^a: $r^a u_a \sim \overset{\circ\circ}{u}{}^a u_a + \overset{\circ}{u}{}^a \overset{\circ}{u}_a = (\overset{\circ}{u}{}^a u_a)^\circ = 0$ (because of $u^a u_a = -1$) – hence a spacelike vector – which measures the *radiative loss* of the accelerated charge. Such radiative losses are extremely small in non-nuclear situations, as measured by the small time constant $\tilde{\tau}$ in front of the second-order time derivative of the velocity; yet they imply all of observational astrophysics. They may even become sizable in situations of large coherence – like in pulsar radio emission – when many (N) charges radiate in phase (and the emitted power per charge grows as N). The 3-vector equivalent of (3.1) reads: $d(\gamma\boldsymbol{\beta})/dt = (e/m_0c)(\boldsymbol{E} + \boldsymbol{\beta} \times \boldsymbol{B}) + \gamma^{-1}\boldsymbol{r}$.

The Lorentz–Dirac equation is of third order of differentiation in the position vector so that its solutions require a position, a velocity, plus an acceleration as complete initial data, contrary to standard practice. Half of its solutions are of the self-accelerated *runaway* type and must be discarded. Its wanted solutions are automatically obtained when one proceeds iteratively (for small r^a), by replacing its highest-order term by the dominant time derivative of the left-hand side: $\overset{\circ\circ}{u}{}^a = d\overset{\circ}{u}{}^a/d\tau \approx (e/m_0c)\,(\overset{\circ}{F}{}^{ab}u_b + F^{ab}F_{bc}u^c)$. The radiation vanishes if and only if r^a vanishes, which happens for uniform acceleration: $\overset{\circ}{u}{}^a\overset{\circ}{u}{}_a = $ constant implies $r^a\overset{\circ}{u}{}_a = 0$, whence $r^a = 0$ for 2 spacetime dimensions (remember: $r^a u_a = 0$). I.e. not only a uniformly moving charge, but even a uniformly accelerated charge does not radiate, as anticipated above.

For non-uniform acceleration, the *power* radiated by an electron follows as

$$L \overset{e}{=} -m_ec^2 r^\circ/\gamma = (2e^2/3c)\overset{\circ}{u}{}^a\overset{\circ}{u}{}_a(1 - \overset{\circ\circ}{u}{}^\circ/\gamma\overset{\circ}{u}{}^b\overset{\circ}{u}{}_b) , \qquad (3.3)$$

an expression which deviates from the often-encountered textbook expression only by the last term in parentheses which tends to be small (of order γ^{-2}). To leading order, therefore, insertion of (3.1) and expression through 3-d field strengths yields the two versions

$$L \overset{e}{\approx} (\sigma_Tc/4\pi)F^{ab}u_bF_{ac}u^c = (\sigma_Tc/4\pi)\gamma^2[(\boldsymbol{E} + \boldsymbol{\beta} \times \boldsymbol{B})^2 - (\boldsymbol{E} \cdot \boldsymbol{\beta})^2] \qquad (3.4)$$

which are worth memorizing, for multiple applications; they contain the Thomson cross section $\sigma_T := (8\pi/3)\,(e^2/m_ec)^2 = 10^{-24.2}\mathrm{cm}^2$. Note that L scales as $(e^2\gamma/m_0)^2$ for particles of different charge, mass, and speed, hence dominates for electrons of large Lorentz factors.

For the more practically oriented reader, this last (approximate) radiation formula could have been derived, in the NR limit, from the more widely known *Hertz* power of an oscillating *dipole* of moment $\boldsymbol{D} := e\boldsymbol{x}$, with a dot again denoting the ordinary time derivative ($\dot{\boldsymbol{D}} := d\boldsymbol{D}/dt$):

$$L_D = (2/3c^3)\ddot{\boldsymbol{D}}^2 = (2e^2/3c^3)\ddot{\boldsymbol{x}}^2 \approx (2e^2/3c)\overset{\circ}{u}{}^a\overset{\circ}{u}{}_a , \qquad (3.5)$$

which is seen to agree approximately with (3.3). The dipole formula – and its generalization to higher multipoles, scaling as the square of the $(n+1)^{th}$

time derivative of the 2^n-pole moment – often yields a useful expression for calculating antenna powers.

Yet another approach to the radiated power is offered by the quantum picture: Think of an electromagnetically acccecelerated charge as one that *Compton scatters* on plane waves, or photons of energy $h\nu$, with a mean free time $\tau = 1/n\sigma_T v$ between collisions. As the photon is elastically reflected – for low enough inertia – in the rest frame of the electron, its energy is boosted in the lab frame, on average, by the factor $\approx 2\gamma^2$ (see (3.19)). This leads to a Compton-scattered power

$$L \approx 2\gamma^2 h\nu/\tau = \sigma_T c\gamma^2 2nh\nu = (\sigma_T c\gamma^2/4\pi)(E^2 + B^2) , \qquad (3.6)$$

where the photon energy density $nh\nu$ has been replaced by Maxwell's energy density $(E^2+B^2)/8\pi$. When compared with (3.4), all that has to be changed, in application to more general field geometries, is the replacement of this energy density by the expression in square brackets: it is a more complicated quadratic form in the field strengths that controls, in general, the amount of radiation.

In view of their importance, let us discuss a few extreme applications of (3.4). As these four cases we consider (i) a *linear accelerator* ($B_\perp = 0$), (ii) *synchrotron* (or cyclotron) radiation ($E = 0$), (iii) an $E \times B$-*drift* ($\beta = E \times B/B^2$), and (iv) *Compton scattering* ($0 \approx E - \beta \times B$). In these four cases, the quadratic form in square brackets simplifies, respectively, to:

$$[(E + \beta \times B)^2 - (E \cdot \beta)^2] = \{E^2/\gamma_\parallel^2 , \beta^2 B_\perp^2 , 0 , 4B_\perp^2\} ; \qquad (3.7)$$

we have used the intuitive notation: $1 - \beta_\parallel^2 =: \gamma_\parallel^{-2}$, and $B_\perp := | \beta \times B/\beta |$. Note that the radiation by a linear accelerator is strongly suppressed ($\sim \gamma^{-2}$), and that it vanishes for $E \times B$-drifting charges, whereas it booms for charges moving head-on against a wave field. Note also that (3.3) above has told us that the radiation from a constant linear accelerator is not only small but vanishes strictly: we should not have used the simplified formula. The fact that $E \times B$-drifting charges do not radiate may find an important realization in the extended jet sources, Chap. 11, in which the power of extremely relativistic electrons propagates almost loss-free through $\lesssim$ Mpc distances.

Our above sketch of a quantum derivation, leading to (3.6), was not only meant to show consistency with the photon description. It also reminds us that our classical expressions for the radiated power require modification whenever the momentum of the radiated photon approaches that of the emitting charge: The assumption of an elastic reflection in the electron's rest frame breaks down for photon energies $\gamma h\nu \gtrsim m_e c^2$, ($\gamma$ being the electron's Lorentz factor), i.e. for photon frequencies $\nu \gtrsim 10^{20.2}$Hz$/\gamma$ (called γ-rays for $\gamma \approx 1$), where *Thomson scattering* changes into *Klein–Nishina* scattering, of reduced intensity. A similar reduction occurs in strong magnetic fields, for $eB\hbar\gamma \gtrsim m_e^2 c^3$.

Problems

1. What fraction $\Delta W/W$ of the acceleration power of an electron (to final energy $W = \gamma m_e c^2$) is lost to radiation in a) a *linear accelerator* of length l, b) a *ring accelerator* of radius $a = 10^2$m? Assume the accelerating field strength E to be constant, e.g. $E = \text{kV/cm}$.

2. What is the degradation e^{-1}-folding time $\tau := -E/\dot{E}$ of a relativistic electron of Lorentz factor γ which radiates in a transverse magnetic field $B_\perp$ at a) X-ray frequencies: $\nu = 10^{18}$Hz, $B_\perp = 10^4$G, b) visible frequencies: $\nu = 10^{14.2}$Hz, $B_\perp = 10^{-3}$G (*Crab nebula*), and c) radio frequencies: $\nu = 10^9$Hz, $B_\perp = 10^{-4}$G?

3. How large is the maximal voltage $W/e = \int (\vec{\beta} \times \vec{B}) \cdot d\vec{x}$ to be drawn at its *speed-of-light cylinder* from a magnet rotating in vacuum? Put $\beta \lesssim 1$, $B(r) = B(r_*) \, (r_*/r)^3$, $|\Delta x| \lesssim c/\Omega$, and calculate $\gamma = W/m_e c^2$ for (an electron and) $B(r_*) = 10^{12}$G, $r_* \Omega/c \leq \sqrt{\hat{M}/r_*} \approx 0.6$, i.e. for a typical neutron star.

3.2 Frequency Distributions of Single Emitters

The frequency distribution and polarization of electromagnetic radiation emitted by some medium is the superposition of those of its constituent charges, or multipoles. It reflects their *accelerations*, predominantly those of the lightest particles: electrons, as the emitted power scales as m^{-2}. Let us look at several of them in order.

The simplest accelerated motion is the gyration of a charge around a constant magnetic field, as occurs in a *cyclotron*. The *gyro* (or *Larmor*) radius follows from force balance between attractive Lorentz force $e\beta_\perp B$ and centrifugal repulsion $m v_\perp^2/a$, $m := \gamma m_0$ being the relativistic mass, and $v_\perp$ the velocity perpendicular to B. The angular frequency of gyration $\omega_B := v_\perp/a$ thus follows as

$$\omega_B = eB/mc , \tag{3.8}$$

i.e. scales as B/m with varying magnetic-field strength and particle inertia. Circularly polarized waves are emitted, whose sense is dictated by the sign of the charge, which dictates the sign of gyration.

Once the speed of a gyrating charge approaches the speed of light, its radiation changes to *synchrotron radiation*. Its instantaneous beam, or *antenna lobe*, is a narrow cone in the forward direction, of opening angle $\lesssim \gamma^{-1}$, Fig. 3.1a. Consequently, during a full cycle of the charge's spiral motion around a magnetic field line – which will have an inclination angle ϑ w.r.t. the line of sight – an observer sees a moving charge only when it approaches her or him within $\lesssim \gamma^{-1}$, i.e. during a retarded-time interval of order $\Delta t \approx 2P_{ret}/\gamma$,

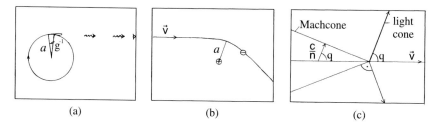

Fig. 3.1. (a) A Charge moving relativistically on a circle, balanced by the Lorentz force of an axial magnetic field, emits synchrotron radiation into a tangential forward cone of opening angle $\approx 1/2\gamma$. A distant observer sees short spikes of radiation, of width $\approx \gamma^{-3}P$, because photons outrun the radiating charges by (only) $1 - \beta \approx 1/2\gamma^2$. **(b)** Electric Dipole radiation is emitted when an electron is deflected, during its thermal motion, by a positive ion. **(c)** Čerenkov radiation is emitted by a charge moving at superluminal speed v through a medium of lower-than-c phase velocity c/n, $(n > 1)$: $\beta > 1/n$. Instantaneous emissions have wave normals at an angle θ w.r.t. $\boldsymbol{v}$ given by $\cos\theta = 1/\beta n$; they fill the Mach cone dragged along by the charge

$(P_{ret} = P/\sin\vartheta)$, during which the emitted photons outrun the relativistic charge by as small a time fraction as $(c-v)/c = 1-\beta$, so that $\Delta t \approx P/\gamma^3 \sin\vartheta$ holds, and the emitted intensity $I(t/t_s)$ has the shape of a fence of period P_{ret}, and pillar width Δt, see Fig. 3.2. Its Fourier transform $\tilde{I}(\nu/\nu_s)$ gives the spectrum; it contains all the overtones of the (ER) gyro frequency $\nu_{B,ret}$ from 1 to some γ^3, and can be approximated (smoothed) – for each of the two elliptical polarization modes – by an Airy integral containing Bessel functions of fractional index $k/3$, which after Wallis [1958] has the convenient approximation (exact for one mode, approximate for the other):

$$\tilde{I}(x) = x \int_x^\infty K_{5/3}(y)dy \approx 1.78\, x^{0.3}\, e^{-x} \, , \, x := \nu/\nu_s \, , \quad (3.9)$$

with ν_s near the upper turnover. I.e. the spectrum starts as a power law of index 0.3, near $\nu = \nu_B$, and has its upper synchrotron-cutoff angular frequency ν_s near the γ^2 overtone of the NR gyro frequency:

$$\nu_s = 3\gamma^2 eB_\perp/4\pi m_0 c = 10^{6.6} Hz \, (\gamma^2 B_\perp)_0(m_e/m_0) \, , \quad B_\perp := B\sin\vartheta \, , \quad (3.10)$$

see Fig. 3.2.

After these two most frequent non-thermal radiation processes, we now turn to *thermal* emissions. The elementary act is the near encounter, or scattering of one particle by another, via electric multipole forces. In a thermal plasma, the dominant radiation mode is dipole radiation by an electron performing monopole collision (*free-free radiation*) on an ion, Fig. 3.1b. Note that

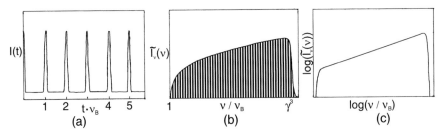

Fig. 3.2. (a) Lightcurve $I(t)$ reaching a distant observer from a gyrating relativistic charge. (b) Fourier transform $\tilde{I}_\nu(\nu)$ of the lightcurve shown in (a). (c) Double-logarithmic presentation of $\tilde{I}_\nu(\nu)$: the (synchrotron) spectrum has slope $\alpha := \partial \ln \tilde{I}_\nu / \partial \ln \nu = 0.3$ between its cutoffs, at the relativistic gyration frequency ν_B and at $\gamma^3 \nu_B$

any scattering of identical monopoles – like electrons on electrons – has a vanishing dipole contribution because their dipole moment $\boldsymbol{D} := e_1\boldsymbol{x}_1 + e_2\boldsymbol{x}_2$ is proportional to their center-of-mass location $m_1\boldsymbol{x}_1 + m_2\boldsymbol{x}_2$, hence constant. For charges with different e/m ratios, the lighter one orbits along a hyperbola around the heavier one whose semi-major axis a follows from the balance of electric attraction and centrifugal repulsion: $mv^2/a = Ze^2/a^2$, whence the typical scattering radius for a *Coulomb* collision:

$$a_{Cb} = Ze^2/mv^2 \tag{3.11}$$

and cross section $\sigma_{Cb} = \pi a_{Cb}^2$. We are again interested in the typical angular velocity $\omega = v/a = mv^3/Ze^2 \approx (kT/\hbar)(\beta/\alpha Z)$, with $\beta := (v/c)_{thermal}$, $\alpha := e^2/\hbar c = 1/137$, and because of $mv^2/2 = 3kT/2$, and notice that the emitted photon energy $\hbar\omega$ is a small fraction of the charge's kinetic energy for $\beta \ll \alpha Z/3$ – which holds for temperatures below some 10^8K. The corresponding characteristic frequency $\nu_c = \omega_c/2\pi$ of thermal scattering is worth noting:

$$\nu_c = (kT/h)(3\beta/\alpha Z) = 10^{14}\text{Hz } T_4^{3/2}/Z \; . \tag{3.12}$$

It is somewhat lower than the Wien frequency $\nu_W \lesssim 3kT/h$ which characterizes the peak of a Planckian distribution at temperature T; such slightly higher frequencies require slightly tighter collisions with higher collision velocities at formation and/or subsequent thermal upscatter.

So far we have restricted all considerations to the motions of the emitting particles, which are transferred undistorted to the emitted photons if the emissions take place in vacuum; but vacuum is always an approximation. In general, the surrounding medium – solid, liquid, gaseous, or plasma – has a *refractive index* n different from unity:

$$n = c/v_{ph} = ck/\omega \; , \tag{3.13}$$

where $v_{ph} := \omega/k$ is the *phase velocity* of a perturbation of wave vector $k = 2\pi/\lambda$ and angular frequency ω. Phase velocities are often smaller than c, for optically thick media, but are larger than c for non-magnetised plasmas: For a one-component, magnetised plasma, textbooks find

$$n_{pl}^2 = 1 - \frac{1}{(\omega/\omega_{pl})^2 + \epsilon\omega/(\omega_{pl}^2/\omega_B)} \quad , \quad \epsilon = 0, \pm 1 \ , \tag{3.14}$$

in which the plasma frequency

$$\omega_{pl}/2\pi =: \nu_{pl} = \sqrt{ne^2/\pi m_e} = 10^{13.9}\text{Hz } n_{20}^{1/2} \ , \tag{3.15}$$

is the frequency at which its negative charges (electrons) oscillate at resonance w.r.t. their positive partners. In any case, an emission is significantly influenced by the embedding medium as soon as n^2 deviates from unity by $\gtrsim (\gamma - 1)^{-1}$ in magnitude, where γ is the Lorentz factor of the emitting particle; see (3.17).

For later use, we also require the *Lorentz–Lorenz* formula for the corresponding dielectric constant ϵ due to the bound electronic resonances in a neutral medium. It reads

$$\frac{\epsilon - 1}{\epsilon + 2} = \frac{-\omega_{pl}^2}{3\omega^2}\sum_\beta \frac{b_\beta}{1 - (\omega_\beta/\omega)^2 + i\delta_\beta\omega} =: -\frac{\hat{a}}{3} \tag{3.16}$$

with complex oscillator strengths b_β of magnitude near unity, and (small) damping constants δ_β at the resonance frequencies ν_β. For a vanishing electric conductivity σ, and a permeability μ, the corresponding refractive index n equals $\sqrt{\epsilon\mu}$. When $\hat{a}$ is small and $\mu = 1$, we have $n^2 \approx 1 - \hat{a}$, an expression similar to (3.14).

Before we come to a discussion of these influences of a non-trivial medium, it may be worth mentioning that a superluminal phase velocity does not conflict with the relativistic doctrine that no signal can propagate faster than at the speed of light. The propagation speed of a wave packet can be shown to be given by its *group velocity* $v_{gr} := d\omega/dk$ if well defined, i.e. if the packet keeps identity and shape, but loses this property in frequency domains of anomalous dispersion. Signal propagation can be recognised as being equivalent to the propagation of transitions from zero to finite amplitude whose Fourier transform involves the short end of the wavelength range. *Signal speed* is therefore measured by the *front velocity* $v_{front} := \lim_{k\to\infty} (\omega/k)$ which is never larger than c [Krotscheck and Kundt, 1978].

Returning to synchrotron radiation in non-vacuum, Schwinger et al. [1976] have shown that its emitted intensity $\tilde{I}(x)$ keeps its spectral shape – on transition to $n \neq 1$ – when the normalised frequency $x := \nu/\nu_s$, (3.10), is replaced by the more general expression

$$x = \nu/\nu_s[1 + (1 - n^2)(\gamma - 1)]^{3/2} \ , \tag{3.17}$$

where emission – between cutoffs – requires x to be < 1; see also Crusius and Schlickeiser [1988]. In particular, emission is exponentially suppressed for Lorentz factors below the *Razin–Lorentz* factor $\gamma_R := \nu_{pl}/\nu_B$, and for frequencies below the *Razin–Tsytovich* frequency $\nu_R := (\gamma\nu_{pl}^3/\nu_B)^{1/2}$. This has applications to X-ray detectors, to pulsars, and to the active nuclei of galaxies.

As a special case of radiation in a non-trivial medium, let us focus on *Čerenkov radiation*. Čerenkov radiation is emitted by charges moving at superluminal speed, which requires $n\beta$ to be > 1, see Fig. 3.1c. A well-known example is the blue light visible in the water tank bathing a nuclear reactor. The condition for the formation of a Mach cone is $\cos\theta = 1/n\beta < 1$, which implies an upper cutoff frequency via $\beta n(\nu_{\max}) = 1$ (as n tends to 1 for $\nu \to \infty$). A lower cutoff results from the constraint that a non-magnetised plasma shorts out electric fields below its resonant *plasma frequency*, (3.15); note that terrestrial air has (number) density $n = 10^{19.4}\text{cm}^{-3}$, i.e. it would oscillate at IR frequencies if turned into a plasma. Between its lower and upper cutoff, Čerenkov radiation is distributed as a power law of index $\alpha \lesssim 1$: $I_\nu(\nu) \sim \nu^\alpha$.

Here we have changed notation, in dropping the Fourier-transform tilde on the *spectral intensity* I_ν, and in expressing its frequency dependence by a lower index, as in partial differentiation:

$$I = \int I_\nu d\nu = \int \nu I_\nu \, d\ln\nu \ . \tag{3.18}$$

Note that the *spectral intensity* $I_\nu := \partial I/\partial\nu$ has a different dimension from the intensity I – viz. energy per area – whereas 'I' can mostly be well approximated by the maximum of its integrand νI_ν w.r.t. $\ln\nu$, which shares its dimension. For this reason, plots of νI_ν vs ν are preferred over those of I_ν vs ν.

Returning once more to Čerenkov radiation, an additional constraint on its frequency range should be mentioned: that quantum theory forbids its frequencies to largely exceed the radiating charge's *de Broglie* frequency, $\nu_{\max} \lesssim 10c/\lambda_{dB}$, ($\lambda_{dB} = h/mv$); this condition is, however, often implied by the above constraint of superluminality. Another remark concerns a result elaborated by Schwinger et al. [1976]: that synchrotron and Čerenkov radiation should be considered as branches of a *synergic* process, i.e. of a uniform, indivisible radiation process – *synchrotron-Čerenkov* radiation – which can be described by a replacement as in (3.17) for both $n\beta <$ and > 1.

Among the important emission mechanisms in the Universe ranges also *Compton scattering* – the relativistic version of Thomson scattering – of electrons on photons, which tends to be called *inverse Compton* when the photon gains energy in the collision. As already mentioned in the derivation of (3.6), Compton scattering means an almost elastic reflection of a photon by an electron in its rest frame, whose 2-fold boost from and back to the lab frame yields an increase in the photon's frequency from ν to ν_{sc} given approximately by

$$\nu_{sc}/\nu \approx 2\gamma^2(1 - \cos\vartheta)/(1 + \gamma^2\vartheta'^2) , \qquad (3.19)$$

where ϑ , ϑ' are the angles of the incoming, and scattered photon respectively w.r.t. the electron velocity in the lab frame; $\gamma\vartheta' \lesssim 1$. With some skill, this result can also be derived from the conservation of the sum of the two incoming 4-momenta during the collision. For monoenergetic colliders, the upscattered photons form a flat distribution of spectral index between 0 and 1. Distributions of photons are upscattered into similar distributions, higher in frequency by a factor of $\gtrsim \gamma^2$. This mechanism is probably responsible for most of the hardest sources in the Universe, with photon energies above some MeV.

Yet another radiation mechanism deserves mention: *synchro-Compton* radiation, the radiation by charges crossing a *strong wave*. Its name tells us that it shares properties of synchrotron with those of Compton radiation. A wave is called *strong* when test particles are boosted by it towards the speed of light within a small fraction (f^{-1}) of its period, i.e. when the *strength parameter*, or non-linearity parameter f exceeds unity:

$$f := e\,B\,/\,m_e\,c\,\omega_w . \qquad (3.20)$$

Among the terrestrial sources of strong waves are our strongest lasers. They can be used as particle accelerators because a strong wave pushes a test charge in the direction of its motion – independent of its sign – via the Lorentz force acting on the charge's transverse velocity. This behaviour differs from that of a weak wave which causes transient oscillations transverse to its direction of propagation. A (presently achieved) laser power of $10^{15}\,\mathrm{W} = 10^{22}\mathrm{erg/s}$ suffices to boost electrons to Lorentz factors $\gg 1$, though not yet protons. The most prominent emitters of strong (magnetic dipole) waves in the Universe are the rotating, strongly magnetised neutron stars, as their low spin frequencies $\omega/2\pi$ allow for long phase intervals of one-way acceleration, i.e. for large f-values. Such strong spherical waves are thought to post-accelerate their relativistic winds. They may well be the long-sought *booster* to the highest cosmic particle energies.

Whereas charges crossing a weak wave emit Compton radiation, the spectrum in a strong wave results as in synchrotron radiation, with a spectral index of 0.3, and an upper cutoff frequency

$$\nu_s \approx \gamma^2\,f\,\nu_w . \qquad (3.21)$$

Problems

1. Radiation at near-IR frequencies, $\nu \lesssim 10^{14}\mathrm{Hz}$, requires what state parameters if generated by   a) *blackbody* radiation at temperature T (for which

$\nu_{peak} \lesssim 4kT/h$),   b) *synchrotron* radiation by electrons of Lorentz factor γ in a field B,   c) (inverse) *Compton* radiation in a heat pool of (room) temperature $T = 300$ K?

2. What *radiation temperature* T_r – defined by $I_\nu =: 2\pi kT_r/\lambda^2$ – possesses an isotropic source which – like *Sgr A** – emits a power of $L \gtrsim 10^{36}$erg/s at frequencies near $10^{12.5}$Hz from within a radius r of $\lesssim$ AU ($=10^{13.2}$cm)?

3. Calculate the *strength parameter* $f := eB/m_e c\Omega$ for the magnetic dipole wave radiated by the rotating magnet in problem 3.1.3. Assume that at the speed-of-light cylinder, the corotating dipole field changes abruptly (in the higher derivatives) but continuously into the wave field.

3.3 Spectra Emitted by Ensembles

Let us begin by recalling the definitions of a few quantities, like *spectral intensity* I_ν, absorptivity κ, emissivity ϵ_ν, and optical depth τ. $I_\nu(\nu)$ is defined as the differential power dL emitted by a source per differential frequency interval $d\nu$, differential area $d^2\sigma$, into a differential spherical angle $d^2\Omega$:

$$dL =: I_\nu \, d\nu \, \cos\theta \, d^2\Omega \, d^2\sigma \tag{3.22}$$

at an angle θ w.r.t. the area's normal, in which the dimensions of the differentials are expressed in an obvious manner, with e.g.: $d^2\sigma := dxdy$; see Fig. 3.3a. For an isotropic radiation field, its intensity I can be shown to relate to its energy density u via:

$$I_\nu = (c/4\pi) \, u_\nu \, , \tag{3.23}$$

a relation needed when background fluxes are to be related to their reservoirs.

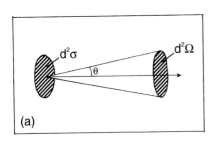
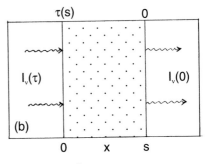

Fig. 3.3. (a) Emission geometry by a surface element $d^2\sigma$ into a small spherical angle $d^2\Omega$, of cone opening angle θ, and: (b) radiation transfer through a slab of thickness s, optical depth $\tau(s)$

The intensity of radiation traversing a non-trivial medium decreases exponentially with path length s, due to absorptive losses described by the *absorption coefficient* $\kappa(\nu)$:

$$\kappa(\nu) := -dI_\nu/I_\nu ds ; \tag{3.24}$$

its dimension, obviously, is 1/length. It is related to the *optical depth* τ of a layer via $d\tau := -\kappa ds$.

More generally, radiation propagating through a non-trivial medium suffers both absorptive and scattering losses as well as emissive gains, the latter described by the *emission coefficient* ϵ_ν whose dimension is power per volume and frequency, so that:

$$dI_\nu/ds = -(\alpha + \sigma)I_\nu + \epsilon_\nu \tag{3.25}$$

describes an unrestricted situation. For local thermal equilibrium (LTE), the gains are the sum of greybody emission αB_ν – with $B_\nu(\nu; T)$ the Planck intensity at temperature T – plus scattering gains σJ_ν, which in many cases can be lumped together into a multiple $(\alpha + \sigma)S_\nu$ of a *source function* $S_\nu(\nu)$. With the more general definition

$$d\tau/ds := -(\alpha + \sigma) \tag{3.26}$$

of the optical depth τ, the *radiation transfer equation* (3.25) thus generalizes to

$$dI_\nu/d\tau = I_\nu - S_\nu . \tag{3.27}$$

This is an inhomogeneous linear differential equation with the general solution

$$-I_\nu e^{-\tau} \big|_0^{\tau(\nu)} = \int_0^{\tau(\nu)} d\tau S_\nu e^{-\tau} \approx S_\nu(1 - e^{-\tau}) \approx \begin{cases} S_\nu \tau , & \tau \ll 1 \\ S_\nu , & \tau \gg 1 \end{cases} , \tag{3.28}$$

the latter for a constant source function S_ν, see Fig. 3.3b. In principle, every signal from a cosmic source is somewhat distorted by the intervening medium and should be reduced according to this equation.

As a frequent application, consider looking through dusty interstellar clouds. *Dust grains* of size l ($\lesssim \mu m$) scatter all wavelengths λ larger than l, in proportion to $(l/\lambda)^4$, hence lose their far-IR transparency increasingly with decreasing λ [Rayleigh scattering, a limiting case of Mie scattering; cf. Lang, 1998].

We are now ready to discuss the spectra radiated by typical cosmic emitters and begin with the most fundamental of all, *Planckian* or *blackbody radiation*. Thermodynamics teaches us that independently of the composition of a source, it emits a universal spectrum if at uniform temperature and sufficiently thick, whose (differential) energy density $u_\nu d\nu$ is the product of the photon number density $dn(\nu) = 8\pi\nu^2 d\nu/c^3$ and their average (zero-rest-mass Bose–Einstein) equilibrium energy $h\nu/(e^{h\nu/kT} - 1)$:

$$u_\nu \, d\nu = \frac{8\pi h\nu^3/c^3}{e^{h\nu/kT} - 1} \, d\nu \ , \tag{3.29}$$

see Fig. 3.4a, in which the Planck spectrum is plotted double logarithmically. Considering that cosmic sources can have spectra which range from low radio frequencies, of order MHz, up to TeV photon energies – i.e. through more than 20 orders of magnitude – a Planckian does not differ much from a delta function.

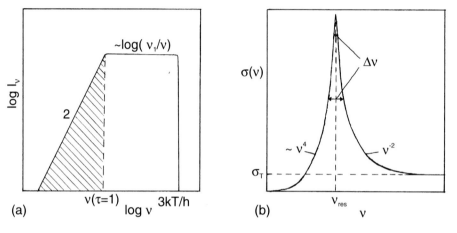

Fig. 3.4. (a) Spectrum of a Thermal Continuum source: $I_\nu(\nu)$ has spectral slope $\alpha = 2$ in the (low-frequency) optically thick regime, and $\alpha \lesssim 0$ above, up to the cutoff near $3kT/h = \nu_u$. (b) Qualitative resonance curve of a (strong) Spectral Line, described by a (weakly) damped harmonic oscillator

The peak of a Planckian distribution – more specifically of its energy density νu_ν – can be obtained from (3.29) by evaluating the maximum of the function $x^4/(e^x - 1)$, or rather of its natural logarithm $4\ln(x/(e^x - 1))$, as a monotonic function of it: $0 = 4/x - 1/(1 - e^{-x})$, whence $x_{max} = 3.92$. This result, $< h\nu > \lesssim 4kT$ – often stated for the maximum of u_ν rather than νu_ν – is known as *Wien's displacement law;* it may be considered the relativistic generalization of the NR thermal equi-distribution law $< mv^2/2 > = 3kT/2$.

How thick must a medium be in order to emit undiluted blackbody radiation? The non-trivial answer to this question reads

$$\tau_{th} = \left(\int n_e^2 ds \right)_{24.8} / T_4^{3/2} \nu_9^2 \ , \tag{3.30}$$

i.e. τ_{th} passes through unity at 1GHz for a plasma temperature of 10^4K when its *emission measure* $EM := \int n_e^2 ds$ passes through $10^{24.8}$cm^{-5}, as e.g. for a column length of 20 kpc at a (large!) mean-squared average density of 10 electrons per cm^3. We confirm that most laboratory conductors are optically thick at GHz (w.r.t. free-free radiation), whereas the Galactic ISM is transparent down to $\lesssim$ 10 MHz. Note that the emission measure enters into the above equation because free-free radiation scales as the rate of binary encounters; in contrast to line absorption which scales as the *dispersion measure* $DM := \int n_e ds$.

An *optically thin* medium emits a thermal spectrum τB_ν, according to (3.28) and (3.29), which is flatter (softer) than a Planckian by a power of two: $I_\nu \sim e^{-h\nu/kT}$, whereby a logarithmic *Gaunt factor* has been ignored; see Fig. 3.4a.

Let us now turn to relativistic-plasma, or *non-thermal* emitters. Most distributions in nature which owe their existence to a large number of processes are *power law*, and so are most of the cosmic energy distributions of radio-emitting relativistic electrons – called *negatons* and *positons* in the book by Jauch and Rohrlich [1955] – whereby charge neutrality requires positive and negative charges to be equally numerous. We describe a power-law distribution by its differential number N_E (w.r.t. energy):

$$N_E dE \sim E^{-g} dE ,\qquad (3.31)$$

and correspondingly the emitted frequency distribution:

$$I_\nu d\nu \sim \nu^\alpha d\nu ,\qquad (3.32)$$

(in which half of the literature defines α with the opposite sign: be aware!). Whenever the emitted frequencies peak at γ^2 times some constant frequency – like in synchrotron-Čerenkov and Compton radiation – and re-absorption can be ignored, there is a simple connection between the two spectral indices g and α. It depends – as limiting cases – on whether or not the observed charge population ages during the observation: An *aging population* transfers its energy to the radiation field, $E_{em} \sim E$, whereas a *non-aging* one realizes $E_{em} \sim E^2$. We thus get for the power: $dL \sim \left\{ \begin{matrix} EN_E dE \\ E^2 N_E dE \end{matrix} \right\} \sim \nu^\alpha d\nu$ emitted by an $\left\{ \begin{matrix} aging \\ non-aging \end{matrix} \right\}$ population, with $\nu_{em} \sim E^2 : \left\{ \begin{matrix} 1-g \\ 2-g \end{matrix} \right\} = 2\alpha + 1$, or:

$$\alpha = -\left\{ \begin{matrix} g/2 \\ (g-1)/2 \end{matrix} \right\} \text{ for } \left\{ \begin{matrix} aging \\ non-aging \end{matrix} \right\}. \qquad (3.33)$$

In words: the spectrum of an – optically thin – aging population is softer, by index $1/2$, than that of a non-aging population. Often a spectrum contains a break in power by $1/2$ at the frequency above which aging takes place. Note that a *white* energy distribution, $g = 2$, radiates a *white* spectrum νI_ν = const.

So far we have considered optically thin populations, for which re-absorptions can be ignored so that the intensities emitted by individual charges add up. A derivation of the *optical depth* τ_{syn} of a synchrotron-emitting layer transcends the scope of this book; as a simplest case, one finds:

$$\tau_{syn} \approx \nu_8^{-3} \, B_{-2}^2 \, D_{17}^{(2)} \quad \text{for} \quad g = 2 \; , \tag{3.34}$$

where $D^{(g)} := \int \gamma^g n_\gamma ds \; \approx \, <\gamma^{g-1}> \int \gamma n_\gamma \, ds$ generalizes the dispersion measure to a relativistic electron population, with $\gamma n_\gamma =: n_{rel}$ measuring the number density of relativistic electrons near the peak of the distribution. For different power laws, the exponents of frequency and magnetic fieldstrength in (3.34) vary with g.

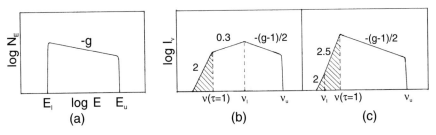

Fig. 3.5. (a) Power-law Electron-Energy distribution, in double-logarithmic representation. (b,c) Non-aging Synchrotron Spectrum, with lower cutoff frequency ν_l {larger, smaller} than the transition frequency $\nu(\tau = 1)$ from optically thick to thin; horizontally plotted is $\log \nu$

For the fairly general case of a *single power-law energy distribution* with sharp lower and upper cutoffs E_l and E_u, it has been shown in the literature that its synchrotron spectrum has a broken-power-law shape, with spectral index 2 in the thick case – at low frequencies, as for Rayleigh–Jeans (thermal) radiation – followed in the thin case by 0.3 below the frequency ν_l corresponding to the low-energy cutoff in the electron-energy distribution, further $-(g-1)/2$ above that frequency up to the upper break frequency ν_u unless aging takes over before, with index $-g/2$; see Fig. 3.5. Here we have skipped over yet another possible power-law interval with exceptionally hard spectral index 2.5 which occurs between the lower break frequency ν_l and the transparency frequency $\nu(\tau = 1)$, if the latter is higher. Equipped with this knowledge, it should be possible to calculate from a measured spectrum the electron populations which have emitted them.

Frequency spectra consist of continuum parts with lines superposed, both in emission and in absorption. *Spectral lines* are due to atomic and/or molecular resonances. As for a harmonic oscillator, their frequency-dependent excitation cross section $\sigma(\nu)$ has a steep maximum at some resonance frequency ν_r which can exceed its non-resonant part (σ_T) by a factor of $\lesssim 10^{15}$:

$$\sigma(\nu) \approx \sigma_T \nu^4 / [(\nu^2 - \nu_r^2)^2 + \nu^2 (\Delta\nu)^2] \,, \tag{3.35}$$

with σ_T = Thomson cross section, and with a typical line width of $\Delta\nu/\nu_r \gtrsim 10^{-7.5}$ at optical frequencies (which can be lowered, via cooling, to $\gtrsim 10^{-12}$), see Fig. 3.4b. In realistic situations, light is emitted by moving atoms, and $\sigma(\nu)$ must be Doppler-averaged over a thermal population, with $\delta\nu/\nu \approx \pm\beta_{th}$ (according to (1.18)). We then get – with $r_e := e^2/m_e c^2$ = classical electron radius –

$$< \sigma >_{th} \approx (\beta_{th}\nu)^{-1} \int_0^\infty \sigma(\nu) d\nu = \frac{f\pi r_e \lambda}{\beta_{th}} = 10^{-12.1} \text{cm}^2 \ f \ \lambda_{-4.5} \ T_4^{-1/2} \tag{3.36}$$

for a line of strength parameter $f \leq 1$, wavelength $\lambda = 10^{-4.5}$cm, emitted by hydrogen at $T = 10^4$K. This averaged cross section is still over 10^{12}-times larger than σ_T for a strong line; it can be memorised as $\lesssim \lambda\lambda_e\alpha/2\beta_{th}$ with $\alpha := e^2/\hbar c$, $\lambda_e := h/m_e c$.

Clearly, a star can grow huge when photographed in the light of a suitable spectral line at which its windzone is optically thick out to some large radius. Note that in a stellar windzone, the radial motion of the emitting atoms, or ions tends to be supersonic, much faster than their thermal velocities, and the Doppler-shifted line cross section becomes highly anisotropic. *Radiation-transfer* calculations therefore require more care for lines than for the continuum.

Problems

1. How many *photons* are there in a volume λ^3 of a *blackbody* radiation of temperature T, where $\lambda \, (= c/\nu)$ is the wavelength at its energetic maximum (defined as the peak of $\nu B_\nu(\nu)$)?

2. For a *power-law distribution* of particle (kinetic) energies E with cutoff energies $E_{l,u}$, $N_E dE \sim \left\{ \begin{smallmatrix} E^{-g} dE \\ 0 \end{smallmatrix} \right\}$ for $\left\{ \begin{smallmatrix} E_l \leq E \leq E_u \\ else \end{smallmatrix} \right\}$, determine the approximate mean energy per particle $\langle E \rangle = U/N$ in the intervals $g \ll 1$, $1 < g < 2$, and $g \gg 2$; $U := \int E N_E dE$, $N := \int N_E dE$.

3. Similarly as in Fig. 1.5, approximate a *galactic disk* as a homogeneous flat box made of ionized hydrogen of (electron) density $n_e = 10^{-1.5}$cm^{-3}, temperature $T = 10^{4.2}$K, height $2H = 10^{2.8}$pc. At what angle ϑ (w.r.t. its perpendicular) does it become *transparent* at (radio) frequency ν?

4. What apparent *radius* $R(\nu)$ does a *star* have in the light of a strong *spectral line* $(f = 1)$ in whose wind flows a mass rate $\dot{M} = 10^{-6}$M$_\odot$/yr at velocity $v = 10^3$km/s, when the line belongs to an ion of (relative) abundance $Z = 10^{-5}$?

5. What *optical depth* τ must a layer of temperature $T_1 \, (\gg T_2)$ have in order to appear as bright as a blackbody of temperature T_2 a) bolometrically, b) at frequencies near the spectral maximum of T_2?

4. Thermal Processes

The Universe is far from thermal equilibrium: its temperature is $\lesssim 3$ K whereas its visible constituents (stars, galaxies) often have surface temperatures near 10^4K. Cosmic clouds can be colder than 10 K in their cores; γ-ray bursts signal local temperatures near 10^{10}K. How fast do these temperatures change? On large spatial scales, heat conduction is unable to compete with *heat radiation*. This chapter is therefore primarily devoted to radiative heat exchange.

4.1 Entropy Balance and Cooling

A cup of coffee takes a few minutes to cool, from hot to drinkable; how long does a cosmic cloud take? A rigorous answer to this simple question involves the *law of entropy*, which reads for a system at local temperature equilibrium (LTE):

$$T(\dot{s} + s\Theta) \stackrel{LTE}{=} \Gamma - \Lambda + D \ . \tag{4.1}$$

Here $s := S/V$ denotes the (volume) density of entropy S, the latter defined as the potential of the integrable differential form $dS := \delta Q/T$ where δQ is the (non-integrable, differential) reversible heat transfer to the system. Entropy is perhaps the least intuitive among all physical quantities. Its dimension is energy divided by temperature, like that of specific heat. In statistical mechanics, entropy can even be defined for any system (far from equilibrium but) described by a particle-distribution function in phase space. It can never shrink for a closed system; hence the smaller the entropy of a system, the farther is the system from its final steady state. Numerically, the entropy of an equilibrium system in units of the Boltzmann constant is usually a little larger than the number of its particles, photons included. Entropy grows during irreversible processes, like escape of perfume from a scent bottle, or like explosions, or feasts.

We return to the last equation: Its left-hand side (LHS) measures the comoving change of s, with a dot denoting the comoving time derivative $d/dt := \partial_t + \vec{v} \cdot \nabla$, and with $\Theta := \nabla \cdot \vec{v}$ denoting the *expansion scalar* which measures differential volume changes of the streaming fluid medium. Θ is negative for contractions. The RHS is the sum of *radiative* plus *conductive gains* $\Gamma := (dE/d^3xdt)_{gain}$, minus *radiative* plus *conductive losses* $\Lambda := (dE/d^3xdt)_{loss}$, and the (non-negative) rate of *dissipation* plus *heat deposit* $D := \eta \, \sigma^{ij}\partial_i v_j + \partial_i q^i$ inside the system; $\sigma_{ij} = \partial_{(i}v_{j)}$ = shear tensor, see (5.15) and (6.4). The equation is intuitive once we understand that every conserved density, like particle-number density n, satisfies the conservation law

$$\dot{n} + n\Theta = 0 \qquad (4.2)$$

as a consequence of *Gauss's integral theorem*

$$\dot{V} = \int\!\!\int \boldsymbol{v} \cdot d^2\boldsymbol{x} = \int\!\!\int\!\!\int \nabla \cdot \boldsymbol{v} \; d^3x = \; <\Theta> V \; , \qquad (4.3)$$

because of $nV = N = $ const., so that $\dot{n}/n = -\dot{V}/V$; see Fig. 4.1. Note that in thermo-hydrodynamics, the growth of entropy is described continuously by a differential equation, not discretely by referring to its values at initial and final steady states of comparison processes.

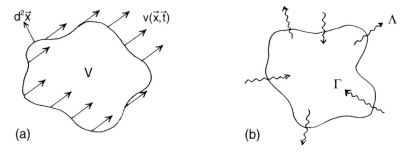

Fig. 4.1. Sketch of (**a**) comoving volume V in a fluid geometry, and (**b**) radiative heat exchange of some fluid domain

Our desired cooling equation follows from (4.1) once we know how the entropy density s depends on the more basic state parameters. For a one-component, non-quantum, non-relativistic system, this expression takes a simple form when use is made of the *thermal de Broglie wavelength* λ_{th} which measures the quantum extent of a particle of mass m in a thermal pool. We have:

$$s \; \overset{NQ,NR}{=} \; nk \; [5/2 - \ln(n\lambda_{th}^3)] \qquad (4.4)$$

with

$$\lambda_{th} := h/\sqrt{2\pi mkT} \; . \qquad (4.5)$$

Note that the argument of the logarithmic term is the volume occupied by a thermalized particle, λ_{th}^3, in units of the average volume per particle n^{-1}; it is < 1 for systems dilute enough to be described by the NQ formula, so that the logarithm is negative, and s is positive. Once particles start overlapping, T must be replaced by the *Fermi temperature* T_F defined by

$$kT_F := E_F = \left\{ \begin{array}{ll} (3\pi^2 n)^{2/3}\hbar^2/2m & , \text{NR} \\ (3\pi^2 n)^{1/3}\hbar c & , \text{ER} \end{array} \right\}. \tag{4.6}$$

In (4.4), the argument of the logarithmic term, $n\lambda_{th}^3$, is constant during quasistatic, adiabatic expansions and contractions, and so is S.

At this point, note what had been anticipated above: *Entropy* per k at equilibrium is equal to the particle number except for the expression in square brackets of (4.4) which varies slowly with density and temperature. This expression makes entropy non-trivial. It implies the non-additivity of entropy when different systems are brought into contact. Yet s/nk varies only between values of order unity – for cold, dense systems, like laboratory devices, or compact stars – and $\lesssim 10^2$ – for hot, dilute systems, like the IGM. In the ultra-relativistic limit, s/nk approaches $\{3.6, 4.2\}$ for $\{$bosons, fermions$\}$. When an old stellar core shrinks from solar size to become a white dwarf, or neutron star, its entropy shrinks monotonically. The irreversibility of this process is reflected by the growth of the entropy of the complete system, core plus emitted radiation.

We are now ready to derive the *cooling equation*, by evaluating (4.1) with (4.4), for simplicity's sake only for a one-component system without dissipation. Its LHS can then be simplified successively as $T(\dot{s} + s\Theta) = Ts\ \dot{}\ (s/n) = -(Ts/[\ ])\ \dot{}\ \ln(n\lambda_{th}^3) = Tnk\ \dot{}\ \ln(T^{3/2}/n) = p\ \dot{}\ \ln(T^{5/2}/p)$, whence the full equation:

$$\frac{d}{dt} \ln\left(\frac{T^{5/2}}{p}\right) = \frac{\Gamma - \Lambda}{p} , \tag{4.7}$$

in which the LHS simplifies further for *constant pressure*, to $(5/2)\dot{T}/T$. On introducing the reduced loss and gain functions $L := \Lambda/n^2$ and $G := \Gamma/n$, which take care of their different scalings with density, due to binary or single interactions, we then get for the temperature e^{-1}-folding time t_{cool}:

$$t_{cool} = -T/\dot{T} = (5/2)kT/Ln = 10^6\text{yr}\ T_4/L_{-24}n_{-1} . \tag{4.8}$$

I.e. for typical interstellar densities $n \lesssim 0.1$ cm^{-3} and a temperature $\lesssim 10^4$K, cooling to cloud temperatures takes longer than a Myr, the precise e-folding time depending on the exact value of the reduced cooling function L, Fig. 4.2. For laboratory devices, on the other hand, cooling of radiatively thin (IR transparent) systems takes small fractions of a second. Of course, the last estimate ignores compensating gains by incoming radiation.

Note that in Fig. 4.2, the *cooling function* L – the power radiated per volume and density squared – has been obtained by adding all emissions from

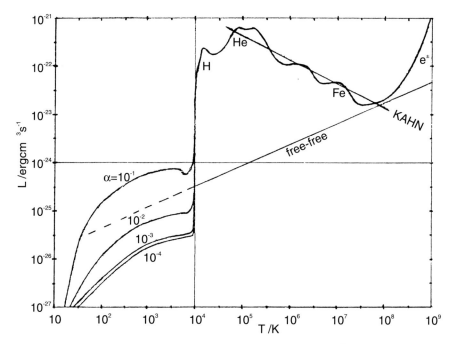

Fig. 4.2. The reduced Cooling Function, $L(T) := \Lambda(T)/n^2$ vs T, in double-logarithmic presentation. For a plasma, free-free radiation is a minimal radiative loss, strongly assisted by line radiation above 10^4K and by pair formation above 10^8K. Below 10^4K, cooling depends sensitively on the degree of ionisation $\alpha := n_e/(n_e + n_i)$. Between 10^5K and 10^8K, L can be approximated by Kahn's power law $L \approx 10^{-21.88} T_6^{-1/2}$ erg cm^{3}s^{-1}

a medium of local Galactic composition, both collisional free-free radiation from its ionized component and the various line radiations. For temperatures lower than 10^4K, the degree of ionization depends strongly on history, and so does cooling.

Problems

1. How large is the *entropy per particle* and Boltzmann constant, S/Nk, for   a) terrestrial air,   b) water,   c) intergalactic plasma ($n \lesssim 10^{-5}$cm^{-3}, $T \gtrsim 10^8$K),   d) pre-supernova 'gas' ($n \gtrsim 10^{36}$cm^{-3}, $T \gtrsim 10^{10}$K),   e) Planck radiation?

2. What *Fermi temperatures* T_F have   a) typical metal electrons,   b) the electrons and protons in the interior of white dwarfs ($\rho \gtrsim 10^6$g cm^{-3}),   c) the neutrons in the deep interior of a neutron star ($\rho \approx 10^{15}$g cm^{-3})?

3. What is the cooling (e^{-1}-folding) time via radiaytion in a cold environment of a) a black ball of specific heat like water ($c = 1\mathrm{cal/g\,K}$, cal $= 10^{7.6}$erg), of (homogeneous) temperature T ($\approx 10^3$K), radius R (≈ 10 cm), b) a young SN filament ($n \gtrsim 10^{10}\mathrm{cm}^{-3}$, $T \lesssim 10^4$K, $d \lesssim 10^{14}$cm, degree of ionization $\alpha \lesssim 0.1$), c) an ionized Galactic HI cloud ($n \approx 0.1$ cm^{-3}, $T \gtrsim 10^4$K)?

4.2 Thermal Equilibria

According to the preceding section, *temperature equilibrium* requires equal losses and gains, $\Lambda = \Gamma$. Such an equilibrium is stable if and only if for a growing temperature, the losses grow faster than the gains: $0 < (d/dT)(\Lambda/\Gamma)$. On multiplication by T and under the assumption that Λ/Γ scales as nT^λ with density and temperature, this condition can be reformulated for its logarithm (as a monotonic function of it), into $0 < (d/d\ln T)\ln(\Lambda/\Gamma) = (d/d\ln T)\ln(nT^\lambda)$. By the chain rule of differentiation: $d/d\ln T = \partial_{\ln T} + (d\ln n/d\ln T)_p\partial_{\ln n}$, whence: $(d/d\ln T)\ln(nT^\lambda) = \lambda - 1$. We thus learn that the radiatively stable temperature intervals are charcterized by the inequality $\lambda > 1$, i.e. by a slope >1 of the cooling function.

Figure 4.2 then tells us that the *radiatively stable temperature intervals* are (i) below 10^2K – cosmic clouds – (ii) near 10^4K – the visible ISM – and (iii) above 10^8K – in crashes or detonations, where $e^\pm$ formation acts to stabilise. The temperature 10^4K is further stabilised by the phase transition of hydrogen from neutral to ionized, similar to terrestrial temperatures which are stabilised in cool climates by the melting of ice and/or snow.

A discussion of Galactic thermal processes, like cloud formation and evaporation, requires not only a knowledge of the cooling function but also of the *heating function* Γ: the available radiative heat input. In order to get rough quantitative estimates, we shall assess successively the average powers expected from supernovae, hot stars, stellar winds, and from the cosmic rays. To this end, the molecular-cloud layer of the Galactic disk is approximated by a cylindrical box of radius $R = 10$ kpc, height $H = 10^2$pc, hence volume $V = 10^{66}$cm^3 $H_{20.5}$, (1.11), and Γ can be expressed as $\Gamma = \Delta E/V\Delta t$.

For *supernovae*, we assume a kinetic energy of $\Delta E = 10^{51}$erg – corresponding to 3 M$_\odot$ ejected at a mean-squared average speed of $10^{3.8}$km/s – and a (controversial) repetition time of $\Delta t = 10$ yr – corresponding to the assumption that the $\lesssim 10$ SNe of the past millennium all went off within a $\lesssim 10\%$ vicinity of the Sun, i.e. that we have only detected one in $\gtrsim 10$ SNe, due to occultation by intervening dark clouds. This SN rate should equal the neutron-star birthrate, of both pulsars and non-pulsars, which I assess equally high, one in ten years, see (2.11). Γ_{SN} thus results as $10^{-23.5}$erg/cm^{3}s.

For heating by stellar radiation, we should only use that fraction of the stars' output which gets reabsorbed in the Disk; let's call it the *stellar UV*. It can be estimated either by assuming that a typical bright star, of some 5 M$_\odot$ at birth, converts some 10% of its hydrogen into elements between helium and iron, thereby liberating $\lesssim 10$ MeV/GeV $= 1\%$ of the rest energy per atom, with a corresponding stellar energy of $\Delta E = 10^{52}$erg of which only 10% are reabsorbed. Again, the birthrate of bright stars is assumed equal to the SN rate, one in 10 yr on average, and Γ_{UV} results equal to the SN heating rate, $10^{-23.5}$erg/cm³s. An alternative way of estimating the stellar UV is to estimate the re-absorbed fraction L_{UV} of the Galactic luminosity, $L_{UV} = 10^{44-1.5}$erg/s, and divide it by the volume V of the re-absorption region: $\Gamma_{UV} = L_{UV}/V = 10^{-23.5}$erg/cm³s.

The power output via *stellar winds* is smaller than radiative – for roughly equal radial momenta – by the velocity ratio $\beta_{wind} = v_{wind}/c \approx 10^{-2.5}$, but is expected to be absorbed 100% so that $\Gamma_{wind} = L_{wind}/V = 10^{44-2.5-66}$erg/cm³s $= 10^{-24.5}$erg/cm³s amounts to only some 10% of the former. Note that heating via winds need not proceed radiatively, rather collisionally via shocks, but must, of course, be encorporated into the thermal gains.

Finally, Galactic matter is heated by the *cosmic rays* during their rare collisions. The corresponding rate Γ_{CR} equals the absorbed energy density ϵu_{CR} during a replenishment time-scale $\Delta t \approx 10^7$yr, $\epsilon \approx 0.1$, whence $\Gamma_{CR} = \epsilon u_{CR}/\Delta t = 10^{-27.5}$erg/cm³s, much smaller than the preceding rates. This mode of heating can, however, even penetrate dark clouds whose cores are radiatively screened.

5. Magnetic Fields

Whereas electric fields tend not to play a dominating role on large spatial scales in the Universe – because free electrons are ubiquitous, ready to shunt them – magnetic fields are observed almost everywhere, in pressure equilibrium, raising the *viscosity* of their anchoring plasma, and making its electric and thermal *conductivities* highly anisotropic. Magnetic *torques* can control the spin history of stars and planets, and the *miscibilities* of plasmas are drastically reduced by transverse magnetic fields in their boundary layers. Decaying magnetic fluxes – so-called *reconnections* – are efficient boosters to relativistic particle energies. These properties of magnetic fields apparently make them the key agents for (i) supporting the *solar corona*, (ii) transporting angular momentum in *disks*, *stars*, and *planets*, (iii) driving *supernova explosions*, (iv) forming (magnetically bandaged plasma) *filaments*, (v) forming *jets*, and for (vi) *boosting* charges to relativistic energies. This chapter will concentrate on the build-up and decay of large-scale magnetic fields.

5.1 Fields and Their Amplification

Magnetic fields are encountered on *Earth* ($\approx$G), on the *Sun* (G to 10 G averaged, $\lesssim$ 3 KG in the dark spots), in the *Galaxy* ($\approx$ 5 µG), in *galaxy clusters* ($\lesssim$ 3 µG), and even in *superclusters* ($\lesssim$ 10 µG), ordered on length scales up to Mpc. In their anchoring domains, they tend to be as strong as permitted by the confining static pressure: $B^2/8\pi \lesssim 2n_e kT$, or even ram pressure: $\lesssim \rho v^2$ (in supersonic situations); which shows that the fields are near saturation. If their pressures were larger than the yield strength of their anchors, they would disperse them and behave like in vacuum: they would decay at the speed of light, with part of their energy radiated to infinity. *Dynamos* are required to keep the fields near saturation. It is an unsolved problem of how the early Universe managed to generate fields ordered on supercluster scales: ordered mass motions should have been in action there.

A good conductor *freezes* its *magnetic flux* $\Phi := \int \int \boldsymbol{B} \cdot d^2\boldsymbol{x}$, i.e. drags it along with its motion, because there is no way in which the field-generating electric currents could transfer their energy to the medium. Flux decay is therefore a phenomenon occurring in non-perfect conductors. Whilst a flux is frozen in, its energy can be raised, or lowered, depending on the mode of distortion of its geometry by the motion. For high conductivity, flux conservation implies

$$B \, A \cos \theta = const. \, , \tag{5.1}$$

where A is the area crossed by the flux, and θ is the inclination angle of this crossing, the angle enclosed by $\boldsymbol{B}$ and $d^2\boldsymbol{x}$. From this equation one infers that under *isotropic compression*, B scales inversely as area, $B \sim r^{-2}$, and that B grows as $1/\cos\theta$ under *shearing*, see Fig. 5.1a.

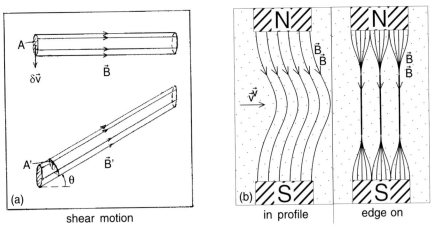

shear motion   in profile   edge on

Fig. 5.1. (a) Amplification of Magnetic-Flux density B by shearing: $B A \cos\theta$ is conserved for growing θ. (b) Splitting of magnetic flux into Tubes (Ropes) by a transverse conductive flow

Field amplification under compression is an important hurdle to *star formation*, so that newborn stars are expected to be maximally magnetised. Field amplification by shearing, on the other hand, may be the dominant way in which *spiral galaxies* – or more generally (conducting, differentially rotating) *disks* – amplify their seed fields to saturation. There is a rich and 'high-brow' literature on galactic dynamos whose goal it is to verify this expectation.

More generally, *magnetic flux* Φ and *vortex strength* $\Psi := \oint \boldsymbol{v} \times d\boldsymbol{x} = \int\int \nabla \times \boldsymbol{v} \cdot d^2\boldsymbol{x}$ of a conducting fluid medium can be both raised and lowered, and transformed into each other according to the integral theorem

$$\frac{d}{dt} \int\int \left(\boldsymbol{B} + \frac{m_i c}{Ze}\boldsymbol{\omega}\right) \cdot d^2\boldsymbol{x} = \int\int \boldsymbol{Q} \cdot d^2\boldsymbol{x} \tag{5.2}$$

in which d/dt is the comoving, or Lagrangean, time derivative, $\boldsymbol{\omega} := \nabla \times \boldsymbol{v}$ the *vorticity* vector occurring as the integrand in the vortex strength Ψ, m_i

and Z are the ionic mass and charge number, and the source vector $\boldsymbol{Q}$ is given by

$$\boldsymbol{Q} = \nu_M \Delta \boldsymbol{B} + \nu \Delta \left(\frac{m_i c}{Ze}\boldsymbol{\omega}\right) + \frac{c}{(Z+1)e}\nabla p \times \nabla n^{-1} , \tag{5.3}$$

with the *magnetic viscosity* $\nu_M := c^2/4\pi\sigma$, $\sigma :=$ electric conductivity, $\Delta :=$ Laplacian operator, and $\nu :=$ kinematic viscosity. This theorem can be straightforwardly derived from the hydrodynamic and Maxwell's equations – by making use of Gauss's and Stokes's integral theorems – for a one-fluid medium of stationary currents ($\partial_t \boldsymbol{j} = 0$) and negligible chemical diffusion. It shows that the vector fields $\boldsymbol{B}$ and $\boldsymbol{\omega}$ satisfy a coupled diffusion equation, of the approximate type $(d/dt - \nu\Delta)\boldsymbol{\omega} = 0$, whose driving term is proportional to the cross product of the gradients of pressure and inverse density. In words: magnetic flux and vortex strength can be transformed into each other. If left alone, they dissipate according to their diffusion equations, in proportion to their viscosity scalars; and moreover, their properly scaled sum can be generated from non-aligned thermal and density variations.

Quantitative evaluation shows that it is difficult to generate significant magnetic fluxes beyond stellar scales l during cosmic epochs: $B \lesssim ctkT/el^2 = 10^{3.4}\mathrm{G}\ t_{17.5}T_4 l_{11}^{-2}$, as was found by Ludwig Biermann in 1950. Once seed fluxes are there, they can be amplified by *disks* at locally conserved fluxes, via squeezing and fusion, and be subsequently expanded by *winds* and *cocoons* to the detected scales.

Problem

1. Inside the speed-of-light cylinder (SLC) of a *magnetised rotator*, the field strength $B(r)$ drops with radial distance as r^{-m}, with $m = \{3, 2\}$ for {vacuum, conducting windzone}. What is the constraint on surface field strength $B(r_*)$ and angular frequency Ω, for the voltage W/e at the SLC to suffice for *pair formation* via electron collision on photons of energy $h\nu \approx \mathrm{eV}$? The condition reads: $W \gtrsim (2m_e c^2)^2/h\nu$. Interesting field strengths are $B(r_*) \gtrsim \{10^4, 10^6, 10^{12}\}\mathrm{G}$ for {stars, white dwarfs, neutron stars}, respectively.

5.2 Conductivity and Flux Decay

Through (5.2) with (5.3) we have learned that magnetic flux decaysdiffusively for a finite conductivity σ, in proportion to $\nu_M \sim \sigma^{-1}$; i.e. the lower the conductivity, the faster the *decay*. For a quantitative evaluation, we need a formula for σ. Realistic cosmic situations will be approached in four steps.

Consider the (1-d) motion of a free electron in the presence of an electric field E. On top of its random thermal motion, the electron will experience a constant acceleration and thereby speed up until it collides with some other particle, typically an ion, deposits its excess momentum, and gets newly accelerated:

$$m_e \ddot{x} = eE \implies \ <\dot{x}> = \dot{x}_{max}/2 = eE\tau/2m_e \ , \tag{5.4}$$

with $\tau :=$ mean free time between collisions $= 1/n\sigma_{Cb}v_{th}$, $n = n_e/Z$, $\sigma_{Cb} =$ Coulomb cross section $\approx (Ze^2/kT)^2$ (see (3.11)), and $v_{th} =$ mean thermal velocity $= \sqrt{3kT/m_e}$, which tends to be large compared with the ordered conduction velocity. The *electric conductivity* σ is defined by $\boldsymbol{j} = n_e e <\dot{\boldsymbol{x}}> =: \sigma\boldsymbol{E}$, whence:

$$\sigma = \omega_e^2\tau/8\pi = \xi(kT)^{3/2}/3m_e^{1/2}Ze^2 = 10^{14}\text{s}^{-1}T_4^{3/2}/Z \ . \tag{5.5}$$

Here, $\omega_e^2 = 4\pi ne^2/m_e$ is the square of the plasma (angular) frequency, see (3.15), and ξ is a tabulated *Gaunt factor*, not too different from unity, which corrects our poor approximation of the Coulomb cross section. The final value, 10^{14}s^{-1} for hydrogen at 10^4K, is the well-tested laboratory conductivity; it is density independent because an increase of conducting electrons is compensated by a reduction in mean free path. It is inferior to the conductivity of copper at room temperature, by a factor of $10^{3.8}$.

Our derivation of the conductivity under laboratory conditions contained the assumption that free electrons move force-free between collisions, with the sole exception of the applied electric field. This assumption deteriorates at low densities when free charges are screened by the ambient plasma, to a radial distance called the *Debye length* λ_D:

$$\lambda_D = \sqrt{kT/4\pi ne^2} \ = 10^{-6.7}\text{cm} \ \sqrt{T_4/n_{19}} \ . \tag{5.6}$$

The Debye – or screening – length shrinks with an increasing density n of the charges, and with decreasing thermal depolarization. At sufficiently low densities, it causes the accelerated electrons to share part of their excess momentum with an ambient screening cloud, i.e. to excite plasma waves such that $\omega_e\tau \approx 8\pi$; and (5.5) yields the new, density-dependent expression

$$\sigma \approx \omega_e \ . \tag{5.7}$$

This low-density formula scales as $\sqrt{n_e}$; for $T = 10^4$K, it meets the high-density branch at $n_e = 10^{19}\text{cm}^{-3}$, slightly below atmospheric density.

Once we try to apply the new static-conductivity law to the magnetic-flux migration on the Sun, we notice that we should not have done so: it predicts flux freezing, whereas the solar dipole reverses every 11.1 yr on average. We have ignored the fact that the solar convection zone is turbulently mixed. In a *turbulent* medium, the ordered motion of the accelerated electrons is permanently offset by disordered bulk motion; the electrons are hampered in trying to transport their charge parallel to $\boldsymbol{E}$. Conductivity in a turbulent medium is controlled by hydrodynamic offset rather than by microphysical braking.

Its quantitative description requires a derivation of the (homogeneous part of the) dynamo equation. Starting from Maxwell's vector equation

$$\nabla \times \boldsymbol{B} = (4\pi/c)\boldsymbol{j} + (1/c)\partial_t\boldsymbol{E} ,$$ (5.8)

dropping the induction-current term by the restriction to *quasi-stationary* situations, introducing the conductivity law

$$\boldsymbol{j} = \sigma(\boldsymbol{E} + \boldsymbol{\beta} \times \boldsymbol{B}) ,$$ (5.9)

and forming the curl, we get (for constant σ)

$$\nabla \times \nabla \times \boldsymbol{B} = (4\pi\sigma/c)[\nabla \times \boldsymbol{E} + \nabla \times (\boldsymbol{\beta} \times \boldsymbol{B})] .$$ (5.10)

Here, the operator on the LHS equals $\nabla div - \Delta$; $\nabla \times \boldsymbol{E} = -(1/c)\partial_t\boldsymbol{B}$; and the curl of the vector product in square brackets can be re-expressed as $\boldsymbol{\beta}(\nabla \cdot \boldsymbol{B})$ $- \boldsymbol{B}(\nabla \cdot \boldsymbol{\beta}) + (\boldsymbol{B} \cdot \nabla)\boldsymbol{\beta} - (\boldsymbol{\beta} \cdot \nabla)\boldsymbol{B}$, of which the first two terms vanish for an incompressible flow. Using $d/dt = \partial_t + c\boldsymbol{\beta} \cdot \nabla$, we arrive at the *dynamo equation*

$$\left\{\frac{d}{dt} - \frac{c^2}{4\pi\sigma}\Delta\right\} \boldsymbol{B} = (\boldsymbol{B} \cdot \nabla)\boldsymbol{v} ,$$ (5.11)

whose driving term on the RHS tends to be used to describe the build-up of magnetic flux from turbulent motion, e.g., in the Sun.

My personal doubts in this well-established procedure of *dynamo theory* are three-fold: (i) Dropping the induction-current term is not legitimate in the presence of narrow flux tubes; (ii) subsequent linearization of the coarse-grained RHS is not permitted in the presence of flux tubes; and (iii) the more rigorous comoving equation (5.2) lacks the equivalent term of the RHS. My scepticism goes further: Turbulent plasmas have a poor conductivity – as we shall derive shortly from the LHS – hence allow magnetic flux to escape, rather than build up. This property has been used for many years to explain the orbital-period gap (between 2 h and 3 h) in the histogram of close binary white dwarfs: the companion's flux escapes as soon as its anchoring surface layers turn turbulent. And it can likewise be used to understand the 22.2-yr magnetic cycle of the Sun.

Let us return to the dynamo equation. For vanishing RHS, it equals the (parabolic) *diffusion equation* which controls the mixing of gases (when $\boldsymbol{B}$ is replaced by n, and $c^2/4\pi\sigma$ by the kinematic viscosity ν), or the approach of thermal equilibrium (when $\boldsymbol{B}$ is replaced by T, and again $c^2/4\pi\sigma$ by ν). Any bump in the initial distribution spreads on the diffusion time scale t_{diff}:

$$t_{diff} = x^2/\nu$$ (5.12)

which grows quadratically with the length scale, as can be seen by replacing derivatives in (5.11) by divisions. In the present case, ν should be replaced by the coefficient $c^2/4\pi\sigma$, and we find the *quasi-static magnetic-flux decay time scale* due to poor conductivity:

$$t_{dec} = 4\pi\sigma(x/c)^2 . \tag{5.13}$$

This decay time grows linearly with the conductivity and quadratically with the dissipation scale.

Statistical mechanics shows the kinematic-viscosity scalar ν to equal $\lambda v_{th}/3$ where λ is a particle's mean free path, or to equal $h v_t/3$ in a turbulent situation where h is a typical scale of the turbulence, somewhat smaller than the largest eddy size, because the turbulent transport is less efficient than a force-free straight-line flight. In order to describe flux decay in turbulent situations, (5.11) suggests that we should replace $c^2/4\pi\sigma$ by ν_{turb}, i.e. introduce a *turbulent conductivity* σ_t by equating $c^2/4\pi\sigma_t$ with $h v_t/3$:

$$\sigma_t \approx c/4h\beta_t , \tag{5.14}$$

in which h is usually equated with the scale height of the considered plasma layer in the prevailing gravity field. Turbulent conductivity therefore depends exclusively on the turbulent length scale and speed.

Problem

1. Below what pressure of a terrestrial vacuum chamber does the (density-independent) *electrical-conductivity* formula for (dense) plasmas lose its applicability?

5.3 Flux Ropes and the Solar-System Magnets

When a plasma moves fast through a magnetised region – as happens frequently in our cosmic surroundings – there is no stationary solution because, as we have seen, field transport progresses as $\sqrt{t}$ and hence falls short, after some finite time, of any motion at constant speed. It is like trying to cross a cage wall made of rubber rods, or like Kippenhahn's specially dressed acrobat jumping from the top of a circus tent down into a magnetised cavity between two strong-current coils. In such situations, the solution of least action has the field compressed into strongly *magnetised ropes,* or *tubes* in pressure balance, interspaced with fieldfree domains, see Fig. 5.1b. Examples are (i) the *solar convection zone* (with emerging flux tubes visible as cool sunspots), (ii) the magnetised solar wind sweeping across the (weakly ionized) *atmosphere of* planet *Venus,* or (iii) various regions of the ISM in which plasma pressure and temperature vary visibly across *magnetised filaments.*

In such situations of a plasma streaming through a grid of flux ropes, the crossing speed is no longer controlled by magnetic forces but rather by the bulk friction of the plasma, similar to Millikan's oil droplets falling through air. But whereas oil droplets can be approximated by spheres, flux ropes should be approximated by cylinders, and instead of Stokes's solution, we require *Oseen*'s solution, e.g. from Landau and Lifshitz VI:

$$\delta F/\delta l \lesssim 4\pi\eta v , \qquad (5.15)$$

which says that the *drag force* δF per length δl on a subsonically moving cylindrical rod is independent of its diameter and proportional to the dynamic viscosity $\eta = \rho\nu$ of the streaming medium as well as to its relative velocity. Under stationary conditions, therefore, a grid of flux ropes is crossed at a speed v determined by equating the above drag force to the force per rope length driving the relative motion, e.g. buoyancy in the solar convection zone.

We are now ready to look at the magnetic structures of the Sun, the Earth, and the other planets, starting with the *Sun*. As already mentioned above, the literature offers different interpretations; see Kundt [1998b]. Most recent work considers a turbulent plasma capable of generating its own magnetic flux whereas it has been argued above that a turbulent plasma is magnetically transmittent, consistent with the period-gap interpretation of white-dwarf binaries. In the first interpretation, the solar convection zone would be a flux generator whereas in the second, it is a flux modulator.

Additional reasons for the *modulator interpretation* are that (ii) the solar surface is very unevenly covered with flux tubes, both latitude-wise (for given hemisphere) and hemisphere-wise, and (iii) the (22.2 ± 2)yr solar (*Hale*) *cycle* can be traced in all the ($\lesssim 90$) lowest magnetic multipole moments, not only in the dipole, as well as in all the surface oscillations, both of even and odd parity, further in the vibrational p-modes, the line radiations ($\Delta L_l/L_l \lesssim 1$), the bolometric luminosity ($\Delta L_\odot/L_\odot \lesssim 10^{-3.3}$), and the wind strengths. There is just too much order for a stochastic interpretation. And (iv) the Hale cycle shows a long-term stability, in the presence of short-term fluctuations, unlike a relaxation oscillator; which has led Ron Bracewell and Robert Dicke to speak of a *flywheel*, or a *chronometer* deep inside the Sun. Such a flywheel may well be a quadrupolar flux frozen into the highly conductive, rigidly rotating radiative interior, whose constituent dipoles make their alternating appearance at the surface once every 11 yr. They are wound up by the differential rotation of the convection zone which varies both with latitude and altitude as well as with the solar cycle, but rise as a result of their lower weight, and of being pulled out into the windzone. See also Fig. 5.2 which predicts the interior rotation profile of the Sun from its 5-min radial surface oscillations.

As a necessary condition for our preferred interpretation, note that (5.5) predicts a diffusive decay time in excess of 10^{11}yr for the ($\gtrsim 10^7$K hot)

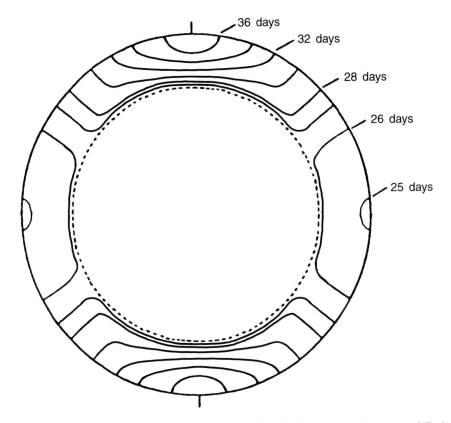

Fig. 5.2. Solar differential Rotation Period $P(r, \vartheta)$ of the convection zone, $0.7 \leq r/R_\odot \leq 1$, inferred from the 5-min surface oscillations. The radiative solar core has been assumed to rotate rigidly, possibly forced by a magnetic torque

radiative solar core, but a transmissivity time of only 10 yr for the *Sun's turbulent convection zone*, of thickness 0.3 R$_\odot$, both by application of (5.14), or of the – less well defined – buoyant-fluxtube interpretation. To me, the sunspots are an indication that the core's flux wants to get out – as in vacuum – but is permanently pushed down by the differential rotation of the plasma layers in the convection zone. Relative streaming of surface layers and emerging flux lead to its splitting into tubes.

What drives the *solar cycle*, and replenishes flux losses? Note that *equatorial superrotation* (of $\Delta\Omega/\Omega \lesssim 0.3$) has the wrong sign to be explained by stirring; it requires agitation. I therefore think that the core of the Sun is spun down magnetically, by friction on its escaping wind, and leaves the bulk of the transmittent convection zone spinning ahead. Both diffusive-flux-loss replenishment and solar-wind driving take their energy partially from the core's rotation, via flux winding, and via centrifugal speeding-up.

Next let us turn to our home planet *Earth*. Its surface field, of strength $\approx$G, is dominantly dipolar, with higher multipoles decreasing exponentially in strength with multipole number ($\lesssim 30$). All moments are time-variable, on time-scales $\gtrsim 10^2$yr; the dipole reverses stochastically, once every $10^{5\pm1.5}$yr. As measured by MAGSAT, at least the first 14 magnetic moments have comparable energies at a depth of (3050 ± 50)km, some 150 km below the surface of the molten metallic core, so that a transiently vanishing dipole can be compensated by a strong quadrupole in screening the surface. Apparently, all the flux is anchored in the molten, highly conductive core whereas the mantle is largely transmittent, with a poor conductivity (of perovskite) of some 10^8s^{-1}. The high-order magnetic anomalies anchored in the mantle drift westward, at a rate of $\lesssim 0.3°$/yr, implying a superrotation of the mantle, like on the Sun. This remarkable fact may tell us that the fluid core is spun down magnetically, on the solar wind, and that at the same time, its otherwise diffusively decaying field ($t_{dec} \approx 10^4$yr) is replenished by flux winding [see Kundt, 1998b]. Flux replenishing may be helped by a loosely coupled solid inner core (at its center), of radius 1020 km, which is thought to spin prograde (superrotate eastward), at fractions of a degree per year.

All the other large bodies in the Solar System have likewise magnetic surface fields, of order $B/G = \{\lesssim 10,\ \lesssim 1,\ \lesssim 1,\ \lesssim 10^{-1},\ \lesssim 10^{-2.3},\ \lesssim 10^{-2.9},\ \lesssim 10^{-3.1};\ 10^{-4\pm1},\ 10^{-2}\}$ for \{*Jupiter, Saturn, Uranus, Neptune, Mercury, Mars, Venus; Moon, Ganymede*\}, measured mostly by unmanned spacecraft; Jupiter's moons Io and Europa appear to have comparable surface fields to Ganymede. A suggestive energy source for the magnetized planets is differential rotation. Besides, alternating super- and subrotation prevails in latitude belts of most of the planets' atmospheres. Kundt and Lüttgens [1998] argue that all these phenomena – with the probable exception of the lunar fields – can be explained by angular-momentum redistribution via magnetic torques in differentially rotating, conducting systems.

A further case of (large) magnetic momentum transfer is the post acceleration of *cometary tails* by the solar wind, see [Kundt, 1998b].

Problem

1. On what time-scale does the *magnetic flux* decay   a) of a cube of copper, of edge length $l = 10$ cm, b) of the solar convection zone (with $\Delta R/R = 0.3$), assuming static conductivity, c) of same convection zone, assuming turbulent conductivity with $h/\Delta R = 10^{-3}$, $v_t = $ km/s, d) of same convection zone, assuming flux tubes rise buoyantly, with drift velocity $v_d = (gA/4\pi\nu)(\delta\rho/\rho)$, $A \approx (10^2$km$)^2$, $\nu = 10^{12}$cm^2/s, $\delta\rho/\rho \approx B^2/8\pi p \approx 10^{-4}$?

6. Disks

Accretion Disks abound in the Universe: (i) in *galaxies*, both spiral and elliptical, (ii) around *forming stars*, (iii) around *planets*, and (iv) around mature and old, degenerate *stars*. They are the result of mass transfer, or mass accretion at large angular-momentum excess. In them, due to non-negligible friction, matter not locked up in bound objects spirals inward at grain-mass dependent rates, allowing for chemical segregation.Accretion onto compact stars leads to large energy releases at high temperatures, both during and after spiral-in, giving rise in particular to the binary X-ray sources. Mass densities in the centers of galactic disks can reach stellar-interior values and hence allow for nuclear burning.

In rare cases, partially ionized gaseous disks can be the result of mass ejection, via a magnetospheric slingshot; examples are the disks around Be stars, and around the outer planets [Kundt and Lüttgens, 1998]. Such disks are called *excretion* disks, or *expulsion* disks.

6.1 Quasi-stationary Accretion Disks

Accretion disks form during contraction, when mass is transferred from an extended reservoir to a more compact one. In them, pressure and viscous forces take care of a smooth redistribution of the angular momentum such that matter ends up in a thin, differentially rotating configuration, mostly as *thin* as a razor blade (in proportion). The fact that disks tend to look thick – like the Milky Way – is partially due to their warping, and partially to their high-kinetic-temperature components, like the older stars; the molecular-cloud layer has a scaleheight of only some 80 pc, at a distance of 8 kpc from the Galactic center.

In order to calculate a disk's half-thickness H, or half-opening angle θ, its confining gravity can be approximated by two constituents: a quasi-spherical acceleration field $\boldsymbol{g}_{sph}$ obeying Coulomb's law for the enclosed mass, assumed at its center – which law is even approximately valid for disk-like source distributions, within a factor of $\lesssim 3$ – plus a 'vertical' component $-2\pi G\sigma$ from the disk's local mass layer, of surface density $\sigma = 2\rho H$, assumed to form an infinite, thin sheet; see problem 1.3.2 and Fig. 6.1a. The sum of the two constituent accelerations has the vertical component

$$-g_\perp \approx \theta GM/r^2 + 2\pi G\sigma =: \zeta\theta GM/r^2 \qquad (6.1)$$

with

$$\zeta = 1 + 2\pi\sigma r^2/\theta M \overset{MW}{\approx} 10^{0.9} , \qquad (6.2)$$

the latter for the *Milky Way* near our Solar System. I.e. the Galactic Disk is heavy, or *self-gravitating* such that above its midplane, the local attraction by its thin disk dominates over the Coulomb-like attraction by the much heavier but more distant masses near its bulge. Note that $\zeta - 1$ can be written as $(2/\theta)(\sigma/ <\sigma>)$, where $<\sigma> := M/\pi r^2$ is the average projected mass per area inside the radius of concern.

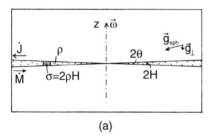 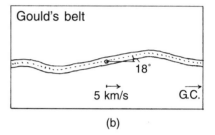

(a)   (b)

Fig. 6.1. (a) Simplified sketch of an Accretion Disk, with its Keplerian angular velocity $\omega(r)$, surface mass density $\sigma(r) = 2\rho(r)H(r)$, half-opening angle $\theta \approx const.$, restoring gravity acceleration $\boldsymbol{g}$, and net flows $\dot{M}$ and $\dot{J}$ of (fluid) mass $M(r)$ and angular momentum $J(r)$. (b) Sketch of Gould's Belt, a likely warp of the Milky-Way disk, of inclination angle $18°$, radial extent 0.5 kpc, and radial speed -5 km/s; GC := Galactic Center

Alternatively, $g_\perp$ can be expressed through its scaleheight H as kT/mH, so that a comparison with (6.1) and $H = \theta r$ yield an estimate of θ:

$$\theta = v_{th} / v_\varphi \sqrt{3\zeta} \overset{MW}{\approx} 10^{6-7.3-0.7} = 10^{-2} ; \qquad (6.3)$$

here, use has been made of the statistical-mechanics relation $kT = mv_{th}^2/3$ between kinetic temperature T and mean-squared thermal velocities v_{th}, as well as of the Kepler law $GM/r = v_\varphi^2$. The disk *half-opening angle* θ has emerged as small as suggested above for the Milky Way; it is likely to be somewhat larger for compact stellar accretion disks, because of their higher kinetic temperatures, yet hardly ever larger than 10^{-1}.

Gaseous matter in disks spirals inward because *angular momentum* is transported outward via friction. In order to quantify this statement, remember that the coefficient of dynamic viscosity η is defined as the shear force per area and velocity gradient for a laminar flow, $-p_{xy}/\partial_y v_x$. More generally, the shear stresses p_{xy} are proportional to the (tensor of) symmetric velocity derivatives:

$$p_{xy} = -\eta(v_{x,y} + v_{y,x}) , \tag{6.4}$$

as becomes intuitive through the constraint that the stresses must vanish for rigid rotation. In cylindrical coordinates, this expression reduces to

$$p_{r\varphi} = -\eta \, r\partial_r\omega = \eta \, \omega \, \beta \tag{6.5}$$

with $\beta := -r\partial_r \ln\omega$ as the exponent of the power-law radial drop of angular velocity ω. The coefficient β equals 3/2 for Keplerian rotation, and 1 for v_φ = const. (typical of outer galactic disks). Now, the frictional momentum-loss density $p_{r\varphi}$ causes a magnitude-wise equal momentum-infall density $\rho v_\varphi v_r$ – matter loses its centrifugal support – and with $\eta = \rho\nu$ we get

$$rv_r = -\beta\nu . \tag{6.6}$$

The radial *infall velocity* v_r can thus be calculated once we know the coefficient ν of kinematic viscosity of the disk substance.

But ν is not an easy quantity to estimate. We know from meteorology that the viscosity of air in turbulent regions can be some 10^9 times larger than in laminar regions, because the entries λ, v_{th} in the microscopic formula $\nu = \lambda v_{th}/3$ have to be replaced by their macroscopic equivalents, *turbulence scaleheight H* and turbulent velocity: $\nu_t \approx Hv_t/3$. Are accretion disks always controlled by *turbulent viscosity*, or are (ordered!) magnetic shear stresses even more important? In face of this uncertainty, Shakura and Sunyaev introduced their famous α-*parameter* through:

$$\nu =: \alpha Hv_t/3 \tag{6.7}$$

with the understanding that α should be $\lesssim 1$, as for $\lesssim$ extreme turbulence. A more rigorous statement is possible, however, by rewriting the above definition into

$$\alpha \approx p_{xy}/p , \tag{6.8}$$

with the help of $p = \rho v_{th}^2/3$ and the approximate equality $v_{th}^2 \approx \theta v_\varphi v_t$ (for sonic turbulence, i.e. $v_{th} \approx v_t \approx \theta v_\varphi$). The RHS of (6.8) is ≤ 1 whenever the *stress tensor* p_{xy} is positive definite, which it is known to be for point-particle systems. Positive definiteness is lost, on the other hand, as soon as magnetic forces gain importance in the stress tensor. In applications to disks, values of α of order 10 or larger are occasionally indicated, suggestive of the importance of magnetic forces. Indeed, in well-sampled galaxies (like NGC 6946), magnetic shear forces can be shown to exert the required differential torque.

For the Milky Way, (6.7) yields a kinematic viscosity of $\nu = \alpha \, 10^{21+6-0.5}$ cm^{2}s^{-1}, and (6.6) an infall velocity v_r of $10^4\alpha$ cm/s. Observed infall velocities of all the young populations in *Gould's belt*, a local warp of our disk of extent some 0.5 kpc and inclination some 18°, are of order 5 km/s, see Fig. 6.1b; they are measured in CO, HI, and on the (young) OB stars, the so-called *K-effect*. If they were typical of what happens at all longitudes, α would have to equal 50. More likely, the K-effect is a strong local fluctuation, and the Galactic spiral-in velocity is presently hidden in the noise.

Once we know v_r, the *mass-infall rate* $\dot{M}_{in}$ follows by integration over a cylindrical strip of area $2\pi r$ times $2H$:

$$\dot{M}_{in} = -4\pi r H \rho v_r = 2\pi\sigma\beta\nu = M\theta^2\omega\xi = v_\varphi c_s^2\chi/G \overset{MW}{\lesssim} 10^{25.8}\text{g/s} , \quad (6.9)$$

the latter $= M_\odot/\text{yr}$, in which formulae the last three versions were obtained by successively using (6.6), (6.7), (6.3), and Kepler's law. The fudge factors $\xi := (2/\sqrt{3})\ \alpha\beta\sqrt{\zeta}\ (\sigma/ < \sigma >)(v_t/v_{th})$ and $\chi := \xi/3\zeta$ are of order unity. According to the four versions of (6.9), we get for our *Milky Way* the mass-infall rates $\dot{M}_{in} / M_\odot\text{yr}^{-1} \overset{MW}{=} \{v_{5.3}, \nu_{27.6}, \xi_{-0.6}, \chi_{-0.7}\}$ which appear somewhat high, meaning that the quantities in curly brackets are all slight overestimates. Note that in (6.9), the Milky Way has been treated as though it was 100% gaseous, instead of only some 10% (by mass); the fudge factors ξ and χ have been adjusted correspondingly.

Once we know the mass rate, we get the *spiral-in time* t_{in} in the form

$$t_{in} = M_{gas}/\dot{M}_{in} = M_{gas}/M\theta^2\omega\xi \overset{MW}{=} 10^{10}\text{yr} / \xi_{0.6} , \quad (6.10)$$

which means that galactic disks are replenished during their lifetimes, hence can change their chemical composition and/or their orientation – in agreement with the fact that galactic disks show chemical gradients, and are often strongly bent, or even inclined at $\lesssim 90°$ w.r.t. the inner disk. For stellar disks, on the other hand, ω is much larger, and θ is not smaller, so that spiral-in times shrink to $\gtrsim$ weeks.

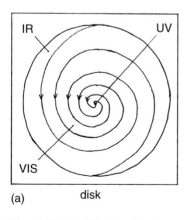
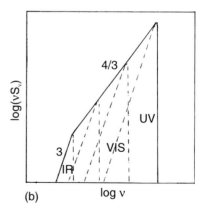

(a)   disk   (b)   log v

Fig. 6.2. (a) Calculated radial Temperature distribution, from the IR to the UV, and (b) emitted Cooling Spectrum of an accretion disk around a white dwarf

Disks heat up as they accrete, because the Keplerian rotation energy $GMm/2r$ increases half as fast as the free-fall energy GMm/r with approach of the bottom of the potential well, at $r = 0$. One half of the *potential energy* is deposited in the disk, via friction, and tends to be thermalized at fixed radius, and radiated away. Under this assumption, the *luminosity* $L = GM\dot{M}_{in}/R$ – with $R :=$ inner accretion radius – is emitted by both sides of the disk, as blackbody radiation, and from $\dot{M}_{in}d(GM/2r) = 4\pi r dr \, \sigma_{SB}T^4$ one finds for the surface temperature T at radius r:

$$T = (LR/8\pi\sigma_{SB}r^3)^{1/4} \, , \tag{6.11}$$

which scales as $r^{-3/4}$, so that the radius-dependent power νS_ν is proportional to $r^2T^4 \sim T^{4/3}$. Wien's displacement law $T \sim \nu$ (for the prevalent blackbody frequency ν) then leads to a predicted *disk spectrum* $\nu S_\nu \sim \nu^{4/3}$, like for monoenergetic synchrotron radiation; see Fig. 6.2. For white dwarfs, these temperatures range from the UV, at the dwarf's surface, to the IR, at the disk's outer edge, and have been confirmed. For galactic disks, on the other hand, the predicted accretion temperatures are cooler than the cosmic background, except in their nuclei.

Problems

1. How long is the typical spiral-in time $t := M/\dot{M}$ of gas molecules in a stationary accretion disk for a) galactic disks of outer radius $r = 10$ kpc, b) stellar disks of outer radius $r \approx R_\odot$ around a central mass $M_c \approx M_\odot$? Use: $t \lesssim 1/\theta^2\omega$, $\theta := h/r \gtrsim 10^{-2}$.

2. Between which frequencies ranges the (power-law) spectrum of a 'naked' accretion disk of inner radius $r_i = 10^9$cm, outer radius $r_o = 10^{11}$cm, around a white dwarf of radius $R = 10^9$cm and accretion luminosity $L = 10^{35}$erg/s?

6.2 Disk Peculiarities

So far, we have restricted considerations to continuous, axisymmetric accretion disks without internal energy input; but there are important exceptions, called *disk instabilities*. Disks can, for instance, form clumps, via self-attraction: Proto-planetary disks form *planets*, protolunar disks form *moons*, and galactic disks form dense clouds which subsequently form *stars*. Our *Solar System* is thought to have formed from a protosolar disk, and so are all the other planetary sytems, around the majority of all other stars. The radius-dependent *chemical diversity* of the Solar-System condensates is due to the preceding action of the proto-planetary *centrifuge*.

Another disk instability is *bar formation*: Potential energy is gained when disk matter at a fixed radius is concentrated at one or more longitudes, under conservation of its angular momentum. A bar through the center is a frequently encountered irregularity, in the inner parts of galaxies, and confuses the kinematics of clouds and stars by violating axial symmetry of the potential; our Milky Way is thought to harbour one as well. *Spiral arms* may be former large-scale bars, wrapped up by differential rotation; they are the preferred sites of cloud and star formation. Bar formation in protostellar disks may explain why stars form in *binaries,* or *multiples: from condensations* on opposite sides from the center.

As in a star, the heat transport in a disk, from its midplane to the surface, can happen either *radiatively*, via photon diffusion, or *convectively*, via buoyancy. These two cooling modes are thought to imply very different viscosities, and α-values, and lead to unsteady mass accretion in disks around compact stars.

A disk can also be cut up and decomposed into massive clumps, or magnetically confined *blades*, by the strong corotating magnetic field of its central compact accretor, a neutron star or white dwarf [Kundt, 1998a]. In that case, the accretion flow continues partially via these blades, in the orbit plane, and partially gets evaporated and ionized and follows the magnetic field lines to the magnetic *polar caps*. A third fraction, injected beyond the corotation radius, can be centrifugally re-ejected, as for *expulsion* disks

There is a limit to the mass rate that can be spherically accreted by the central attractor, known as the *Eddington rate*. This rate is controlled by the average radiation force exerted on a free electron, $L\sigma_T/4\pi r^2 c$ with $L = GM\dot{M}_{in}/R$, which throttles its feeding accretion flow when it exceeds the weight GMm_p/r^2 of its partner proton, yielding the limiting Eddington rate (Fig. 9.2a):

$$\dot{M}_{Edd} = 4\pi m_p Rc/\sigma_T = 10^{18} \mathrm{g\,s}^{-1}\, R_6 \approx 10^{-8} \mathrm{M_\odot yr}^{-1}\, R_6 \;. \qquad (6.12)$$

For a neutron star of radius $R = 10^6$cm, this limiting rate means that it cannot swallow a solar mass, if offered in gaseous form, faster than within some 10^8yr. If its massive binary companion transfers at more than this rate, often some 10^{-5}M$_\odot$/yr – when close enough, and burning fast enough – the transferred matter has nowhere to go but to pile up in the disk, growing in mass towards and beyond several solar masses. Such binaries, containing neutron stars surrounded by massive disks, tend to be interpreted as *black-hole candidates* (BHCs) because the compact component is much heavier than the limiting mass of a neutron star, and often quite dark, and because massive disks are sometimes believed to be unstable. During their formation, their large liberated binding energy gives rise to a bright, *supersoft X-ray source*, of which more than 34 are known in the local group. At later stages, the disk is thought to evolve towards rigid rotation, described by a McLaurin ellipsoid, with a strongly reduced and variable accretion rate dripping from its inner edge [Kundt, 1998a]. Note that an analogous situation does not exist for a white dwarf because its Eddington mass rate is forbiddingly large, some 10^{-5}M$_\odot$/yr.

The maximal accretion luminosity of a compact star can be obtained from (6.12) by multiplication with the available gravitational energy per mass, GM/R; it is called the *Eddington luminosity*, and given by

$$L_{Edd} = 4\pi m_p GMc/\sigma_T = 10^{38.1} \text{erg/s} \; (M/M_\odot) \,. \qquad (6.13)$$

There are at least eight neutron-star binaries in the Galaxy and the Magellanic Clouds which violate this limit, by factors of $\lesssim 10^{1.7}$; these excessively X-ray bright sources are called *super-Eddington*. In them, a heavy accretion diskmay supply the fuel in the form of thin, heavy blades, thus eluding the assumption of spherical accretion.

Eddington's luminosity constraint, (6.13), tends to be likewise applied to non-accreting, luminous stars with the understanding that at a higher luminosity, they would blow themselves apart. In this self-limiting case, however, the constraint can be weakened by the formation of a *porous atmosphere*, i.e. of convective two-phase inhomogeneities [N. Shaviv, 2000: Astrophys. J. *532*, L137–140].

Another disk peculiarity is expected at the *centers of galaxies* where the radial dependence of the mass density $\rho(r) = \sigma(r)/2H \sim M(r)/r^3 \sim v_\varphi^2/r^2$ signals a singular behaviour in proportion to $r^{-2(1-\epsilon)}$ for a rotation-velocity dependence $v_\varphi \sim r^\epsilon$. Galactic rotation curves tend to be flat, $\epsilon \gtrsim 0$. An average galactic mass density of $\rho = 10^{-24}\text{g/cm}^3$ near the Sun thus predicts the density of water, or average density of the Sun, $\rho = \text{g/cm}^3$, at a radial distance of some 10^{10}cm. I.e. the mass density inside galactic disks is expected to grow towards their center, reaching stellar values in their nuclei. Even if the central density should vanish at some fixed time, the center will fill up beyond 10^6 solar masses, for the mass rates calculated above, within little more than a Myr: The galaxy feeds an *active*, nuclear-burning nucleus, a *burning disk* [Kundt, 2000].

Instead, most of my colleagues prefer to think of a supermassive black hole as the central engine of all the *active galactic nuclei* (AGN). They have not convinced me, after more than 20 years. AGN activity requires a refilling engine, with nuclear burning, magnetic reconnections, and explosive ejections of the ashes. Such ejecta may have been mapped in the form of wedge-profiled *emission shells*around isolated galaxies, and may be the enigmatic sources of *QSO absorption lines*, both Lyα and metal lines, see Sect. 1.6. Black-hole formation would require distinctly higher mass concentrations than are ever reached in galactic nuclei, by a factor of 10^2. The quasar phenomenon is a simple consequence of a permanent inward galactic mass flow, at an average of $\lesssim M_\odot/\text{yr}$, which piles up at the center. According to (6.9), a radius-independent mass-flow rate requires the constancy of $\dot{M}_{in} = -2\pi r\sigma v_r \sim r^{2\epsilon}v_r$, realizable at an almost constant flow velocity v_r, because of a rapidly increasing mass density towards the center. See also Chap. 11.

7. Star Formation

Star formation has to overcome four hurdles, the (i) *pressure* (or *Jeans*) hurdle, (ii) *dynamical* (or virial) hurdle, (iii) *angular-momentum* hurdle, and (iv) *magnetic-flux* hurdle. Due to the first two constraints, star formation can only take place wherever enough matter has cooled sufficiently, mainly in spiral arms and galactic nuclei. The other two constraints are met by star formation in redistributing a cloud's excess angular momentum and magnetic flux via disks. Stars are thus born as magnetised rotators with close companions.

7.1 The Four Hurdles

Matter in the Universe tends to be so diluted and kinetically hot that star formation takes place at a moderate rate, primarily inside *galactic condensations*, though modest star formation outside galaxies is indicated, at the 1% level, by detected SNe, PNe, and globular clusters. Let us look at the constraints which the laws of physics impose on star formation, beginning with the constraint set by gas pressure.

(i) A homogeneous, spherical gas cloud is stabilized against gravitational collapse by its *pressure* gradient ∇p which counteracts the inward-pointing *weight* per volume ρg. Approximating $\mid \nabla p \mid$ by the ratio p/r of central pressure to radius, and inserting $p = nkT$, $\rho = mn$, and $g = GM/r^2$ with $M = (4\pi/3)\rho r^3$, we arrive at the critical *Jeans* radius r_{Jeans} below which a cloud is doomed to collapse:

$$r_{Jeans} = \sqrt{3kT/4\pi Gm^2 n} \overset{H}{=} 10^{17.2}\text{cm} \sqrt{T_2/n_6}. \tag{7.1}$$

The *Jeans* mass $M = (4\pi/3)\rho r^3$ residing inside such a homogeneous critical cloud follows as

$$M_{Jeans} = \sqrt{3(kT)^3/4\pi G^3 n}/m^2 \overset{H}{=} 10^{0.9}\text{M}_\odot \sqrt{T_2^3/n_6}, \tag{7.2}$$

and tells us that stars of one solar mass require cooler and denser conditions to form than those normally encountered in galactic HI regions (with $T_2 = 1 = n_1$); they require conditions encountered inside their molecular cores, but also in ram-pressure confined gas pockets at the outer edges of stellar windzones, so-called *Bok globules*. In all other cases, much larger masses are involved in star formation, thousands or millions of $M_\odot$, of which only a small percentage tends to end up in stars during a single cloud collapse because the light emitted by the first generation of massive stars causes reheating, and halts the collapse; see Plate 2.

(ii) But even without reheating by stars, a collapsing large cloud heats up gravitationally, and halts at half its initial radius due to the stabilising kinetic pressure. This *dynamical hurdle* to star formation follows from the *virial theorem* applied to the closed system of its gravitating constituents which reads

$$2E_{kin} + E_{pot} \overset{Cb}{=} 0 \tag{7.3}$$

for particles interacting via Coulomb's r^{-2} force law. Assume that a spherical cloud starts collapsing at rest, from an initial radius R_-, and stops when it has reached its virialised final state, with: $E_{kin} + E_{pot} = E_{pot}/2$. Energy conservation during collapse implies $E_{pot}(R_-) = E_{pot}(R_+)/2$ so that $R_+ = R_-/2$ holds, because of $E_{pot} \sim R^{-1}$. Further collapse requires dissipative interactions with subsequent radiative energy losses. Star formation depends on cooling.

(iii) Dissipation is also required for a removal of excess angular momentum. In order to see this, note that mass conservation during homogeneous collapse implies $\rho r^3 = const.$, and local *angular-momentum* conservation implies $\omega r^2 = const.$ so that

$$\omega_+/\omega_- = (\rho_+/\rho_-)^{2/3} \tag{7.4}$$

holds, and a contraction from interstellar gas densities $\rho_- = 10^{-25}\text{g/cm}^3$ to mean stellar densities of $\rho_+ = \text{g/cm}^3$ implies an increase in spin frequency ω by a factor of $10^{16.7}$. A typical interstellar gas cloud has an angular velocity comparable to that of Galactic rotation, of period $P_- = 10^{15.9}\text{s}$ near the Sun, whereas young stars have observed rotation periods of days, some $P_+ = 10^{5.5}\text{s}$, yielding a ratio of $10^{10.4}$ (only). There is thus an angular-momentum excessby a factor of $10^{6.3}$ to overcome by star formation. It is thought that this excess leads to the formation of proto-stellar disks, and that stars form near their centers, with minimal spin periods of $P_+ = 10^{4.2}\text{s}$. Even then, the spin hurdle amounts to a factor of 10^5.

(iv) Star formation has to overcome yet another hurdle: *magnetic-flux* shedding. Interstellar matter is thought to freeze its magnetic flux on collapse time scales for an ionized fraction down to some 10^{-7}; for yet higher neutrality, so-called *ambipolar diffusion* allows the neutral component to fall through the magnetized, ionized component. Even if the early stages of star formation should deal with exclusively neutral matter, later stages are thought to freeze the flux, and its conservation takes the form $Br^2 = const.$, identical to the conservation of specific angular momentum. Analogously to (7.4), we thus arrive at

$$B_+/B_- = (\rho_+/\rho_-)^{2/3} \,. \tag{7.5}$$

Starting from Galactic magnetic field strengths of $B_- = 5$ µG and ending at stellar surface field strengths $B_+ \lesssim 10^4$G, star formation requires an amplification by a factor of $10^{9.3}$ whereas strict flux conservation offers above factor $10^{16.7}$. Again, an excess by a factor of $10^{6\pm1}$ has to be removed. As a devil may be best expelled by a demon – Germans say "den Teufel mit Belzebub austreiben" – the excess magnetic flux may help a protostellar disk get rid of its excess angular momentum.

Once we understand that disk formation is an essential preceding stage to star formation, we expect young stars to be born with maximal spin, and magnetic flux, and to participate in the *bipolar-flow* phenomenon. Also, knowing of the bar-mode instability of disks, we should not be surprised to find sso many *multiple-star systems*, double, triple, and more. In wide enough binary systems, each of the two components can be the core of an independent solar system. Wide binary systems may subsequently be disrupted, during encounters with other systems, explaining a decreasing binary fraction with the age of a star cluster. It is not clear whether or not our Sun has ever had a (distant) binary companion, and whether planets and moons occur for all masses, or preferentially around low-mass stars.

Problem

1. What inequality must be satisfied by the particle number density n and temperature T of a gravitating hydrogen gas to form stars, or star clusters of mass $M/M_\odot \geq \mu$? ($\mu \approx 1$ for Bok globules, $\mu \gtrsim 10^6$ for globular clusters). What critical temperatures result for   a) $\mu = 1$, $p = 10^{-11}$dyn/cm^2,   b) $\mu = 10^3$, $p = 10^{-12}$dyn/cm^2,   c) $\mu \lesssim 10^9$, $p = 10^{-12}$dyn/cm^2?

8. Stellar Evolution

Stellar masses range (at least) from some 0.07 $M_\odot$ to some 60 $M_\odot$, i.e. through almost three orders of magnitude, with luminosities ranging from $10^{-6}L_\odot$ to almost $10^7 L_\odot$, through almost 13 orders of magnitude. Their radius and surface temperature tend to grow with their mass, and even more so their luminosity, as $L \sim M^{3\pm2}$. All stars burn hydrogen (partially) to helium; the hotter ones further to C, N, O, or even all the way to iron, thereby liberating $\lesssim 10$ MeV per nucleon $= 1\%$ of their rest energy. Burning hydrogen to helium is controlled by a weak nuclear reaction and hence proceeds at the slow, steady *main-sequence* rate, in contrast to more advanced burning which leads to rapid expansions, along the *giant branch* in the *Hertzsprung–Russell* diagram, and to oscillatory instabilities. Whenever *radiative* heat transport from the burning core to the cooling surface falls short of the needs, *convective*heat transport takes over, which can be radius dependent, and lead to chemical mixing. Deviations from spherical structure arise through the stars' spin, through the presence of binary companions with mass transfer, and through the companions' passing to a late stage of evolution, via a (super-) nova explosion.

During their lifetimes, stars can lose significant fractions of their mass via *winds*, increasingly so with increasing mass.

In the presence of excellent and detailed books on stellar structure and evolution – like that by Kippenhahn and Weigert [1990] – this chapter will be particularly cursory.

8.1 Semi-empirical Laws of Stellar Evolution

It is not easy to measure the mass of a distant star unless the latter has a near companion, so that Kepler's lawsapply; observations are therefore more reliably represented as functions of a star's luminosity and radiation temperature than of its mass. Figure 1.2 shows the histogram of an ensemble of nearby Galactic stars as a function of their *luminosity*, $dN/d\log L$ versus $\log L$, both linearly and logarithmically, the latter in order to cover the rare wings of the distribution, which are based on the local group of galaxies. Most stars have luminosities between $10^{-3}L_\odot$ and $1L_\odot$, but stellar luminosities (of black dwarfs [= cold white dwarfs], brown dwarfs [= supermassive planets], or blue supergiants) can be as low as $10^{-6}L_\odot$, or as high as $10^7 L_\odot$, the Eddington luminosity of an object exceeding $10^2 M_\odot$, see (6.13) and the cautioning remarks there: the Eddington limit can overestimate the mass. A conversion from luminosities to masses tends to assume $L \sim M^3$.

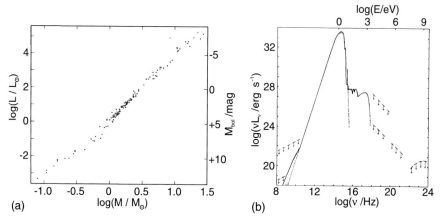

Fig. 8.1. (**a**) Observed Mass-Luminosity relation for 118 binary stars, between $10^{-1} M_\odot$ and $10^{1.5} M_\odot$, and $10^{-3} L_\odot$ and $10^5 L_\odot$. (**b**) Measured Spectrum of the Sun, $\log(\nu L_\nu)$ vs $\log \nu$, from low radio frequencies to $\gtrsim$GeV energies. A best-fit blackbody of $T_{eff} = 10^{3.76}$K is indicated dotted wherever it deviates noticeably. (Fraunhofer) absorption lines prevail softward from $10^{15.3}$Hz (a frequency near the peak, on the Wien branch) whereas emission lines prevail hardward thereof. Besides lines, there are steady excesses, both at radio frequencies and at $\lesssim 10^{6.5}$K, from the inner windzone (corona). Both radio and X-ray emissions tend to flare at various levels, in a correlated manner: $L(\text{soft X})/L(\lesssim 10 \text{ GHz}) = 10^{5.5 \pm 0.5}$ – probably caused by relativistically hot local reconnections in the corona – i.e. are strongly and jointly variable; their culminating levels are represented by *broken lines with vertical arrows pointing downward*

For an empirical handle on the *mass-luminosity* relation, Fig. 8.1a collects data on nearby Galactic binaries at not too close separations (to avoid strong mass exchange), so that Kepler's law allows a measurement of their masses. This figure reveals a systematic dependence which fluctuates smoothly around $L \sim M^3$, between $L \sim M$ and $L \sim M^{5.6}$, and indicates an approach to the Eddington limitation ($L_{Edd} \sim M$) at the high-mass end. For a coarse estimate, note that a star's potential fuel scales as its mass so that $M \sim \int L dt \sim L t$, whence $t \sim M^{-2}$; the more massive a star, the shorter its lifetime. The literature sometimes deals directly with the distribution of (new-born) stellar masses, the so-called *initial-mass function* (IMF), for which Ed Salpeter found the approximate power law $M\dot{N}_M \sim M^{-1.35}$ between cutoffs, ($\dot{N}(M)$ = mass-dependent birthrate).

For a quantitative, numerical treatment of a quasistatic, spherically sym-metric star, one requires (i) an *equation of state* $\rho(p, T)$ (which must incor-porate radiation pressure for hot stars), (ii) a nuclear *burning rate* $\epsilon(p, T)$, and (iii) an *opacity* $\kappa(p, T)$. In addition, the conservation laws of (iv) *mass*, (v) *momentum* (hydrostatic equilibrium, force balance), and (vi) *energy* yield three further relations between the fundamental variables ρ, p, T, M, and L as functions of r which can be conveniently solved for $\{r, p, T, L\} = \{r, p, T, L\}(M)$, i.e. for the listed four functions of the radius-dependent enclosed mass $M = M(r)$. In this program, further use has to be made of the (vii) *heat-flow* law which ignores conduction (as a slow process), but uses *radiative* transport (photon diffusion) in competition with *convective* transport as soon as the radial temperature profile exceeds a critical slope. Note that convective cooling occurs in your kitchen when you heat up your milk, or water too quickly.

Such numerical programs have been run by various groups, see Kip-penhahn and Weigert [1990], with the result that e.g. our *Sun* has central parameters $\{\rho_c, T_c, p_c\} = \{10^{2.0} \text{g/cm}^3, 10^{7.1} \text{K}, 10^{17.3} \text{dyn/cm}^2\}$ during its main-sequence stage (of burning hydrogen to helium). Like all stars below 1.5 M$_\odot$, it should have a *radiatively cooled core* – of relative radius 0.7 – and a *convectively cooled envelope*. Above 1.5 M$_\odot$, this structure is thought to reverse, with the result of frozen, corotating magnetic fields in the radiative envelope (seen, e.g., in the Ap stars), a pattern which should change to fully convective at the high-mass end. If white dwarfs and neutron stars can be born with almost maximal spin, the stellar cores of their progenitor stars must not have been spun down to synchronous rotation with their surfaces; this appears to be consistent with magnetic braking being absent in convective zones.

At the same time, the numerical programs find a *lower mass limit* to igniting hydrogen around $M_H \gtrsim 0.07$ M$_\odot$, a lower limit to (ever) igniting helium around $M_{He} \gtrsim 0.5$ M$_\odot$, and correspondingly $M_C \gtrsim 7$ M$_\odot$, $M_{Fe} \gtrsim 8$ M$_\odot$ for carbon and iron. I.e. stars require a minimum mass of some $10^{32.1}$g. Above 8 M$_\odot$, they are thought to exhaust the nuclear fuel in their cores right up to iron, the most tightly bound chemical element; and above some 60 M$_\odot$, they have been found to explode completely, though observational estimates repeatedly propose even higher values, in conflict with the above luminosity function (which requires supermassive stars to be extremely rare). Burning the elements beyond hydrogen happens on much shorter time scales (than the main-sequence one), which leads to higher ('earlier') temperatures, larger (*giant*) radii, and to oscillatory swelling of the atmosphere because of a temperature-dependent opacity. We thus understand the evolutionary migration of a star through the luminosity-vs-1/temperature (*Hertzsprung–Russell*) plane, with the oscillatory Cepheïd and RR Lyrae stages near the highest-luminosity stripe which connects to the (late) degenerate stages of (non-burning) white dwarfs and neutron stars, via their formation inside planetary nebulae and supernovae; see Fig. 8.2. Empirically, stellar surface temperatures used to decrease through the *spectral classes* O, B, A, F, G, K, M, R, N, S, to be remembered by 'Oh be a fine girl (guy) kiss me right now - smack', which have meanwhile changed into O, B, A, F, G, K, M, L, T, to be remembered by '... kiss my lips - top'.

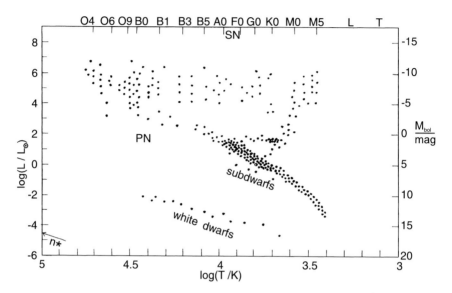

Fig. 8.2. Hertzsprung–Russell diagram of the stars, $\log L$ vs $\log T_{eff}^{-1}$. Earlier presentations tended to use the spectral classes O to M, recently extended to L and T, each subnumbered from 0 to 8, or the spectral colour B-V instead of the effective temperature T. The density of inserted stars has been chosen quasi-logarithmically, in proportion to their detected frequencies in the sky. Most stars appear to the left of the Sun, on the main sequence, whose approximate steepness $\sim T^{-4}$ reflects slowly varying stellar radii. There would be at least as many stars to the right of the Sun (at $\log(T/\mathrm{K}) = 3.76$) on the main sequence were they not so dim (and hence difficult to detect). The dimmest among them are the brown dwarfs: supermassive planets whose cores get hot enough to burn deuterium. Below the main sequence is a group of (metal-poor) subdwarfs, again of slope -4, and distinctly below the branch of cooling white dwarfs. The corresponding branch of (detected) cooling neutron stars lies to the left of the diagram; some dozen of them are known, at surface temperatures of $10^{6\pm0.2}\mathrm{K}$. Above the main sequence are the giants, most frequently the red-clump ones (of absolute magnitude $M \lesssim 0$ mag), as well as the supergiants, ranging all the way up to $10^7 \mathrm{L}_\odot$. Also inserted are the planetary nebulae (PN) – thought to be illuminated by a forming, still much more extended white dwarf – and the (short-lived) supernovae (SN) whose brightest ones fall above the diagram

Stellar *spectra* are dominated by a Planckian at their surface temperature. In addition, a dense windzone may add a power-law plus emission-line excess at frequencies both below and above the spectral peak, and an opaque disk – accretion or excretion – may add another power-law excess (of spectral index $\partial \ln S_\nu / \partial \ln \nu = 0.3$) at low frequencies. On the other hand, a cool chromosphere will remove a number of (*Fraunhofer*) absorption lines below the spectral maximum whereas a hot corona will add emission lines above it. Moreover, small reconnection spots at the surface can add greybody excesses from UV through X-ray to γ-ray frequencies; see Fig. 8.1b. A well-sampled stellar spectrum can therefore tell us various properties of a star: its size, temperature, motions, atmospheric plus windzone state and composition, disk and near surroundings. In many cases, the spectrum will contain contributions from an unresolved secondary companion. We have by no means exhausted the full richness of stellar spectra.

8.2 Modifications due to Spin and Binarity

A static star should be spherically symmetric, forced by its restoring gravity, but real stars *rotate, oscillate*, and *flare*. As argued above, stars are expected to be born with an almost maximal spin frequency (marginally allowed by equatorial mass shedding), near the center of their protostellar disk. What modifications will result?

From the force-balance equation plus equation of state mentioned in the last section, it is easy to convince oneself that to first order, all the stellar parameters should have parallel gradients, including the energy potential $\Omega(r, \omega)$ (which generalizes $M(r)$ for differential rotation at ω) and chemical abundances μ_j, i.e. should have identical constancy surfaces (of axial symmetry: *Poincaré* or *von Zeipel* theorem). All stars should therefore have onion-skin structure.

This first-order result is, however, not rigorous: *Eddington* and *Vogt* have shown that for differential rotation, the heat-flow vector is not divergence-free. Slow *meridional circulations* are implied (via inhomogeneous cooling) which cause a certain amount of mixing. Besides, radial *dredging* will occur due to unstable chemical layering. Even isolated stars can have complex structures.

But stars are seldom isolated. Whenever a star has a sufficiently near stellar companion, *mass exchange* is expected to take place during their evolution. In this process, the more massive star will evolve faster, and transfer part of its wind losses to the originally less-massive partner. If it is massive enough to end its life with a (super-) nova explosion, the mass ratio is likely to revert, and a second epoch of (reverse) mass transfer may ensue during which the compact companion will assemble an accretion disk around it. Clearly, the chemical constitutions of the stellar surface layers will be modified by these mass transfers, and care must be taken in the evaluation of the spectra, and in age estimates.

8.3 Stellar Atmospheres and Windzones

Stars have no well-defined surfaces. Rather, their *atmospheres* are the smooth continuations of their interiors, with a quasi-exponential radial falloff in density. If isothermal, their atmospheres would be unbounded, i.e. would touch the next star, or cosmic cloud. Instead, all stars are found to be surrounded by *windzones* in which their atmospheres escape supersonically at radial distances beyond some critical (Alfvén) distance, until they are sufficiently diluted to be stalled by their circumstellar medium (CSM). The temperature in the windzone would drop adiabatically, $T \sim \rho^{\kappa-1} \sim r^{-4/3}$ (for $\kappa = 5/3$), if there were not simultaneous heating mechanisms at work, like (i) overtaking collisions of faster ejecta, (ii) magnetic reconnections, and (iii) starlight absorptions which keep the windzone significantly warmer than adiabatic.

Stellar mass losses via winds range from $\lesssim 10^{-14} M_\odot/\text{yr}$ to $\gtrsim 10^{-4} M_\odot/\text{yr}$, increasing with the mass and burning rate of the star, whereby the radial momentum flow in the wind can exceed that of the star's radiation by $\lesssim 30$-fold, in particular for (He-rich) Wolf-Rayet stars. What forces propel the wind? An obvious candidate is radiation pressure. But relativistic flows have low momenta; more important are *buoyancy* forces near the surface in combination with *centrifugal* forces, whose pulling gets important near the outer edge of the – magnetically controlled – corotation zone.

9. Degenerate Stars

When massive stars have exhausted the nuclear fuel in their cores, the latter collapse and squeeze their matter to densities of electron degeneracy, or even neutron degeneracy, whereby the liberated binding energy causes a nova-, or supernova-explosion. The stellar remnants formed at the centers of such explosions are compact, non-burning, degenerate stars, *white dwarfs* or *neutron stars*. Their deep potential wells (under accretion) and often high spins and strong magnetic surface fields give rise to all sorts of high-energy processes, among them *pair creation, relativistic ejection, hard radiation,* and *jet formation.* Compact binary stars can assemble compact *accretion disks* around them whose masses may grow comparable in the case of neutron stars (with their low Eddington mass rate), and lead to *supersoft X-ray sources, super-Eddington* accretion, and to the appearance of *black-hole candidates.* For a more comprehensive discussion see [Kundt, 1998a].

Theory also allows for (stellar-mass) *black holes*, but hurdles make their formation rate low. Some of them may hide among the millions of unidentified, steadily glowing soft X-ray sources.

9.1 White Dwarfs and Neutron Stars

Once the hot core of a massive star has exhausted its nuclear storage, it will cool and contract under its own gravity, and under the overburden of the matter lying on top. *Core collapse* will be slowed by adiabatic heating, by centrifugal forces (for significant angular momentum) and possibly by magnetic pressures, but will continue until enhanced (degeneracy) pressures lead to a new equilibrium.

The equilibrium masses and sizes of degenerate stars have been determined in problem 6 of Sect. 1.3, based on the degeneracy pressure (1.13) of {electrons, neutrons} for {white dwarfs, neutron stars}. Their (NR) *Fermi temperatures* $T_F = E_F/k$ follow from (4.6) as $T_F/K = \{10^{9.5} n_{30}^{2/3}, 10^{12.4} n_{39}^{2/3}\}$. They are much higher than the temperatures $T_{form} \approx 10^{-1.5} GMm/Rk \approx \{10^{7.5}, 10^{10.5}\}$K expected at formation due to thermalized gravity, whereby their radii R have been assumed of order $\{10^9, 10^6\}$cm. Consequently, white dwarfs and neutron stars obey low-temperature physics, to be assessed quantum-mechanically. In particular, their heat capacities $\partial_T u$ are lower than for ordinary matter by factors of order T/T_F, so that they cool (or reheat) correspondingly faster. And from an approximate equation of state $p \sim \rho^{5/3}$ one infers the mass-radius relation $R \sim M^{-1/3}$, i.e. a shrinking with increasing mass.

In principle, ages of degenerate stars should be determinable from their (declining) *surface temperatures*. Standard heat conduction theory predicts very high internal conductivities, hence almost isothermal interiors, and a temperature drop of order 10^{-2} through a thin outermost *skin* of non-degenerate and weakly degenerate constitution which measures in meters for neutron stars. *Interior temperatures* T_i would thus exceed surface temperatures T_s by a factor of 10^2. This simple picture requires a number of modifications. First of all note that the plasma frequency (3.15) inside neutron stars largely exceeds expected Wien frequencies so that their interiors should be dark. At high interior temperatures, above $10^{8.5}$K, heat losses from inside neutron stars are therefore thought to take place predominantly via *neutrinos*, whose (minimal) fluxes scale as T^8 until the Planck power sets a limit (implied by Fermi statistics), at $10^{10.5}$K, beyond which flux dependences level off to $\sim T^4$. Neutrino cooling may thus dominate for some first 10^4yr after a neutron star's birth unless there are convective losses, via volcanoes, which expose the hot interior to the outside world and dominate cooling much sooner, possibly right from the beginning.– Only for a handful of nearest neutron stars has one seen thermal radiation from their surface, at $10^{6\pm0.2}$K.

Some of the properties of degenerate stars, thought to be confirmed or suggested by the observations, are summarized in the following Table:

	White Dwarf	Neutron Star	
R/cm	10^9	$10^{6\pm0.3}$	
$M/M_\odot$	≤ 1.4	1.35 ± 0.2	
T_F/K	$10^{9.5}$	$10^{12.5}$	
T_{form}/K	$10^{7.5}$	$10^{10.5}$	(9.1)
T_s/K	$\lesssim 10^5$	$\lesssim 10^{6\pm0.2}$	
T_s/T_i	$\gtrsim 10^{-2}$	$\gtrsim 10^{-2}$	
I/gcm^2	10^{51}	10^{45}	
$P_{\min}/s$	$10^{1.5}$	10^{-3}	
B_s/G	$\lesssim 10^{8.5}$	$10^{12.5\pm1.2}$	

Here, a precise measurement of the *radius* R of a neutron star would be easier within Newton's theory than it is within Einstein's, because of compensating effects. *Mass* determinations, of course, rely on binarity. *Moment of inertia* I are well constrained by the spindown power $L_{sd} = -I\Omega\dot{\Omega}$ of isolated stars. Minimum permitted *spin periods* $P_{\min}$ follow from equatorial force balance,

$$P \geq 2\pi\sqrt{R^3/GM} = 10^{-3.3}\text{s } R_6^{3/2} \tag{9.2}$$

and are almost realized in extreme cases. And *surface magnetic fields* B_s, whilst often well measured spectroscopically for white dwarfs, with a large range probably due to progenitors of quite different masses, are often unknown for neutron stars except when measured via a cyclotron line (at X-rays) and its higher harmonics. An uncertainty in the case of pulsars arises because spindown involves only the polar transverse dipole component; higher multipole components can be vastly stronger. Polar transverse dipole components $B_\perp$ range between $10^{8.2}$G and $10^{13.7}$G, see Fig. 9.1.

Important for an understanding of (core-collapse) supernovae is an estimate of the involved energetics. For the gravitational energy of a static star approximated by an adiabatic compression law $p \sim \rho^\kappa$, Landau and Lifshitz find within the Newtonian approximation: $E_{grav} = -(GM^2/R)$ $3(\kappa - 1)/(5\kappa - 6)$, and for the net *binding energy*:

$$E_{bdg} = E_{grav} + E_{therm} = -\frac{3\kappa - 4}{5\kappa - 6}\frac{GM^2}{R} \ . \tag{9.3}$$

This formula shows that (bound) stars want κ to exceed $4/3$; extreme relativistic degeneracy leads to collapse. General-relativistic estimates have yielded an (uncertain) maximal *stable neutron-star mass* of order 3 $M_\odot$. Unless this mass is immediately provided during a neutron star's formation, the Eddington limit (6.12) may delay its supply by $10^{8.5}$yr or more, thereby delaying, or even preventing black-hole formation.

A fluid star's *magnetic dipole moment* is dynamically unstable, towards a splitting of the dipole parallel to its axis and relative rotation of the two halves such that the dipole is transformed into a (non-axisymmetric) quadrupole. In the case of a neutron star, this instability can be avoided if during the preceding supernova explosion, core collapse leads to an enhancement of the surrounding toroidal field via differential rotation and forms, so to speak, a *magnetic bandage* around the dipole. Such a stabilising bandage creates strong higher multipoles of odd order and is likely to be universally present, more strongly so in the faster-spinning so-called ms neutron stars (if born fast, which I consider obligatory).

Neutron stars are observed in different modes: (i) as (radio) *pulsars* if isolated ($N > 10^3$), (ii) as bright *X-ray sources*, or accretors if accreting mass from a nearby companion ($N > 10^{2.3}$), or (iii) indirectly, as *ejectors* (like SS 433), perhaps as *cosmic-ray* generators and/or *γ-ray bursters* when neither isolated nor steadily accreting, or after having passed their (statistical) age limit as a pulsar, of some $10^{6.4}$yr. Most abundant is mode (iii).

Mode (i) – pulsars – are *broadband*, highly *polarized* lighthouses, with spectral peaks at γ-ray energies, probably with structured (spiky) fan beams whose irregularities are observed through systematic *drifting, subpulse-* and *micro-structure*, whose pulse repetitions can range among the most accurate clocks in the Universe, of accuracy down to 10^{-15}. This rotational stability is often disturbed, by discrete *glitches* ($\Delta P/P \lesssim 10^{-5.3}$) as well as by (temporally unresolved) *spindown noise* both of which may be due to a loose coupling of the (ionized) star to its neutral, superfluid components. The noisiness decreases with a decreasing $\dot{P}$, and beats our best clocks at the small-$\dot{P}$ end. Approximate *pulsar distances d* follow routinely from their dispersion measure $DM := \int n_e ds$, via

$$d = DM \, / \, <n_e> \, , \qquad (9.4)$$

in which the dispersion measure is found from the frequency-dependent delay $\Delta t = d(1/v_{gr} - 1/c)$ of pulse arrivals via

$$\Delta t \approx (e^2/2\pi mc\nu^2)DM \ = 10^{-0.91}\text{s} \, (DM)_{20}/\nu_9^2 \, , \qquad (9.5)$$

see (3.14), and in which the path-dependent mean electron density $<n_e> \approx 10^{-1.55}\text{cm}^{-3}$ requires a reliable Galactic model for higher accuracy; $DM = 10^{20}\text{cm}^{-2} = 10^{1.5}\text{cm}^{-3}\text{pc}$ corresponds to a distance of kpc. A plot of number versus distance shows that our pulsar catalogues become *incomplete* already beyond 0.1 kpc.

Mode (ii) of neutron stars makes its appearance in various ways, as *pulsing* and non-pulsing X-ray sources of high or low system mass which often *burst, flicker, precess,* and *form jets,* are *quasi-periodic, transient, supersoft,* and/or occasionally even *super-Eddington,* with or without *black-hole* candidacy, depending on the mass and constitution of the accretion disk. (For the Eddington mass rate see (6.12) and Fig. 9.2a). Being mostly binary, mode (ii) lends itself to mass determinations; and X-ray spectra allow estimates of the stars' aspect area, and surface magnetic-field strength (via cyclotron lines). Mode (iii) of possible neutron-star appearance will be subject of Chap. 10, and Sect. 13.2 (on SS 433); see also problems 3.1.3, 3.2.3, and 5.1.1.

Many of the details of *pulsars* are still poorly understood, such as (j) their *magnetosphere* structures (bandaged dipoles? Fig. 9.1), (jj) the *work function* of their surfaces at the polar caps (which vanishes in the presence of a pair-plasma corona, generated by electron bombardment), (jjj) the origin of the extreme *brightness temperatures* of their (coherent) radio pulses, reaching $10^{30\pm2}$K in the peaks (of micro-structure and in giant pulses; stimulated small-pitch-angle synchro-curvature radiation, using a modified gyro resonance?), their (jv) radiation at all higher than radio frequencies, most likely incoherent, their (v) high degrees of *polarization,* both circular and linear, (vj) intensity *fluctuations* from pulse to pulse, whose histograms fall into some five different categories, (vjj) 3-d *beam shapes* (or *antenna patterns,* composed of huge numbers of narrow spikes?), (vjjj) modes of *formation* (via a SN explosion in all cases, including the ms pulsars?), and (jx) modes of *extinction*; do the ms pulsars have comparable ages? Even their (x) *spindown ages* $t_{sd} := P/2\dot{P}$ have been occasionally questioned as upper bounds on their true ages, refutably so to my mind: spindown obeys $(P^2)^{\cdot} = 16\pi^2\mu_\perp^2/3c^3 I \approx const.$ within a few per cent (due to a loosely coupled neutral superfluid); the positive initial period $P(t_0)$ is uncertain.

Fig. 9.1. Calculated Bandaged Dipole, which has been proposed as a plausible model for a pulsar magnetosphere in [Ph.D. thesis of Hsiang-Kuang Chang, Bonn, 1994]. An enlargement of the near zone is shown in the *lower right*

Correspondingly, *accretors* pose the problems of their modes of accretion, revealed by the (j) X-ray lightcurve regularities and irregularities known as *quasi-periodic oscillations* (QPO), and occasional good and interlinked periods between 10^2ms and ms, by their (jj) *hard spectra* which have in cases peaked above MeV (for black-hole candidates), their (jjj) occasional *super-Eddington* intensities, and (jv) occasional *jet-formation* capabilities. Sometimes, a rather weak ($\ll 10^{11}$G), or superstrong ($> 10^{14}$G) magnetic surface field has been postulated where instead, the mode of accretion and/or mass in the disk may have been the discriminating parameter. In my understanding [Kundt, 1998a], *X-ray pulsing* results when a significant fraction of the accreting material gets (evaporated and) ionized long before reaching the stellar surface, slides down along magnetic field lines, and lands on the polar caps, Fig. 9.2b. *Unpulsed* X-ray sources result when massive blades reach an equatorial belt unevaporated, cutting their orbits through the corotating magnetosphere as heavy, flux-repelling superconductors. Bursts of type {I, II} are thought to result from {nuclear explosions, gravitational infall} of accreted material. A self-gravitating, rigidly rotating (McLaurin) disk is required to explain the long waiting intervals ($\lesssim$ centuries) between outbursts of the *transient* X-ray sources, during whose formation the source will appear as a bright, *supersoft* source (20 eV to 60 eV) and/or as a *super-Eddington* source;it makes the encircled neutron star look heavy (≈ 7 M$_\odot$), i.e. characterizes a *black-hole candidate*.

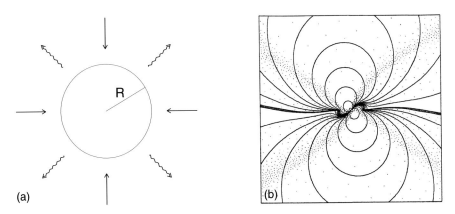

(a)   (b)

Fig. 9.2. (a) Sketch of the Eddington Constraint on a compact accretor: in the limit, the weight of the attracted protons is balanced by the outgoing steady-state radiation pressure on their accompanying electrons, see (6.12). (b) Snapshot of the expected torqued inner Accretion Disk and Magnetosphere around a compact, magnetized rotator which is observed as a variable X-ray source

In parallel with binary neutron stars, there is the zoo of *cataclysmic variables*, viz. binary white dwarfs with their likely *nova cycle*, lasting some $10^{4\pm0.5}$yr, passing through various stages of *recurrent* and *dwarf nova* behaviour including outbursts and superoutbursts with seeming transient orbit lengthening/shortening by $(4 \pm 4)\%$ in the *superhumps* for orbital periods below/above 3 h($M/0.5$M$_\odot$), ($M := M_1 + M_2 =$ total mass in the system), whose attempted explanation by a periodically growing and discharging mass of the accretion disk (between 10^{-10}M$_\odot$ and 10^{-5}M$_\odot$) may lead the right way. Here the superhumps may be caused by illuminated material orbiting outside/inside the Roche lobe. Comparisons of the neutron-star and white-dwarf accretors look rewarding and should certainly be pursued. Their ultimate understanding is difficult as long as they cannot be resolved by our best telescopes.

Problem

1. What fraction of the *binding energy* $E_{bdg} = E_{grav} + E_{therm}$ is stored *elastically* during the *compression* of a homogeneous gas ball   a) in the isothermal non-degenerate case,   b) in the NR degenerate case ($\kappa = 5/3$),

c) in the ER degenerate case ($\kappa = 4/3$)? $E_{grav} = -G \int M(r)\, dM(r)/r = -3 \int p dV$, $E_{therm} = \left\{ {1 \atop 1/(\kappa-1)} \right\} \int p dV$ for $\left\{ {\text{non-degeneracy} \atop p \sim \rho^\kappa} \right\}$.

9.2 Black Holes

There is a maximum mass to a cold, degenerate star set by General Relativity, at $(3 \pm 1)M_\odot$, due to the fact that pressure is not weightless so that for high enough densities, even infinite pressure gradients cannot support a star against its self-attraction; see Problem 1.3.6. Beyond this maximum mass, a celestial body collapses unhaltably and forms a *black hole*, i.e. a concentration of mass which attracts other objects, like ordinary stars, but whose substance is no longer visible from outside because its strong binding forbids even photons to escape: Space inside a black hole contracts so fast that even the local light cones are convected inward. These results from Einstein's theory are among our best physical knowledge. Not equally clear is the height of the *hurdles* that nature has provided to their formation, like centrifugal barriers (in SN explosions), like the Eddington limit (in neutron-star accretion), and like nuclear detonations (at the centers of galactic disks).

Black-hole spacetimes were found mathematically in the form of Schwarz-schild'sspherically symmetric vacuum solutions exterior to a structure-less static assemblage of matter, and of the axially symmetric Kerr–Newman solutions exterior to a spinning and charged source. Much more difficult has been the task of proving that this 3-parameter set of solutions is the complete set of future-asymptotically stationary and predictable vacuum solutions with a smooth future event horizon, i.e. that 'a black hole has no hair': Mass M, spin J, and charge Q determine all higher *multipole moments* M_n and Q_n of a black hole via

$$M_n = M a^n \ , \quad Q_n = Q a^n \ \text{ for } n \geq 0 \ , \tag{9.6}$$

where $ac := J/M$ is the hole's specific angular momentum. The M_n are the effective multipole moments of the {mass, spin} distribution for {even, odd} n, and the Q_n are correspondingly the hole's electric and magnetic moments. During black-hole formation, all initial deviations of the higher multipole moments from the above sequences are thought to be radiated to infinity. Remarkably, a charged, spinning black hole has the same gyromagnetic ratio $Q_1/M_1 = Q/M$ as the free electron.

A complete proof of this *no-hair theorem* does not yet seem to exist, but Heusler's [1996] thesis comes close to it; a much shorter, intuitive version is contained in [Kundt, 1972]. The theorem lies at the heart of (global) General Theory of Relativity. But let us return to the properties of black holes: Their *Schwarzschild radius* R_S is defined through the area A of their surface of no return to the outside world as

$$R_S := \sqrt{A/4\pi} = 2GM/c^2 = 3\,\text{km } M/M_\odot \ . \tag{9.7}$$

In Newtonian language, the free-fall speed of an infalling test particle reaches the speed of light when crossing the hole's horizon. Consequently, a black hole's binding energy is of order of its rest energy. Neutron stars are only some three times less compact, i.e. are prime progenitors for solar-mass black holes, did not the Eddington constraint (of quasi-spherical accretion, limited to $\leq 10^{-8}M_\odot/\text{yr}$) require a massive ($>3\ M_\odot$) and long-lived ($>10^{8.3}\text{yr}$) mass donator, a combination that is marginally at variance with a stellar donator, and even with a (rigidly rotating) massive-disk donator.

At Newtonian approach, the critical *mass density* ρ_{crit} of a black-hole progenitor decreases with its mass as M^{-2}:

$$\rho_{crit} = \rho_N (7\ M_\odot/M)^2 \ , \tag{9.8}$$

where ρ_N equals nuclear density $= 10^{14.6}\text{g/cm}^3$; for galactic masses, ρ_{crit} corresponds to a high terrestrial vacuum. This formula shows that we should distinguish between *mini*, *midi*, and *maxi* black holes, with masses between the *Planck* mass $\sqrt{\hbar c/G} = 10^{-5}\text{g}$, the *Hawking* mass $\hbar c/Gm_\pi = 10^{15}\text{g}$, the *Chandrasekhar* mass $(\hbar c/G)^{3/2}/m_p^2 = 10^{34}\text{g}$, and the cosmic mass (inside the past lightcone) $(\hbar c/G)^2/m_\pi^3 = 10^{55}\text{g}$. Mini black holes, between 10^{-5}g and 10^{15}g, could only exist as fossil remnants of the early Universe (but have so far evaded detection, see below); for midi black holes there is no formation path in sight; but maxi black holes are the ones which have conquered the headlines of all journals, both for stellar masses, and for supermassive ones ($\lesssim 10^9 M_\odot$); their formation densities are below nuclear, hence not prohibitive according to first principles.

Not all combinations of the three fundamental black-hole parameters are permitted: As expected from *centrifugal* instability and *electrostatic repulsion*, there is the fundamental constraint

$$c^2 J^2/GM^2 + Q^2 < GM^2 \tag{9.9}$$

which forbids excessive spin rates and charges. A spinning black hole can transfer part of its rotational energy to the outside world, and thereby lose part of its mass, but not more than the excess over its *irreducible mass*

$$M_{irred} = M[1 - (cJ/GM^2)^2 - Q^2/GM^2]^{1/2} \ . \tag{9.10}$$

Such spinning holes would act as giant grindstones, spewing matter tangentially at high velocities for considerable times.

A somewhat speculative marriage of GR with field quantization suggests the definition of a mass-dependent *black-hole temperature*

$$T = \hbar c^3/8\pi GMk = 10^{-7.1}\text{K}(M_\odot/M) = 10^{12.2}\text{K}\ M_{14}^{-1} \ , \tag{9.11}$$

corresponding to a typical wavelength $\lambda = c/\nu = 2\pi^2 R_S$ of a hole's evaporation radiation (with $h\nu \approx 4kT$) which does not resolve the horizon, i.e. for which the black-hole interior is point-like. This temperature is so low that present-day stellar-mass holes are still perfect absorbers of the 3 K background radiation, but mini holes below 10^{14}g would deradiate on the time scale

$$t_{dec} = 10^{10}\text{yr } M_{14}^3 , \tag{9.12}$$

with a final detonation at γ-ray energies which would trigger off a radio burst. Such radio flashes have never been detected, at the level of 10^{-14} times the critical mass density (1.21) in mini holes. For all practical purposes, therefore, mini black holes have not formed in the early Universe.

How about stellar-mass holes? They can make their appearance in the sky by accreting ambient matter which heats up and radiates before being swallowed, predominantly at soft X-ray energies. As a black hole has non-varying multipole moments, any radiation and variability would have to result from unsteady accretion. Over 45 *black-hole candidates* have been proposed during the past 30 years from the class of binary X-ray sources, both high-mass and low-mass – among them Cyg X-1 and A0620-00 – on account of their large mass functions, absence of strict periodicities, and absence of type-I bursts (understood as nuclear detonations at neutron-star surfaces). To me, all of them look like neutron stars surrounded by massive (≈ 5 M$_\odot$) accretion disks, because of their often hard spectra (up into the γ-ray range), highly structured, fluctuating lightcurves saturating at $L_{Edd}(1\text{M}_\odot)$ during outbursts, line-luminous windzones, occasional jet-formation and/or super-Eddington behaviour, and because of their indistinguishable further properties, as a class, from all the established neutron-star binaries [Kundt, 1998a; also Sect. 13.5]. They just fill the gap between the high-mass and low-mass compact binary systems.

And the postulated *supermassive black holes* at the centers of (all the active) galaxies? They were once believed to be required for energy reasons. But nuclear burning (of H to Fe) is almost as efficient a lamp as black-hole accretion, yielding a guaranteed $\lesssim 1\%$ of the rest energy, and is suggested by the fact that (i) the quasar BLR spectra show $\gtrsim 10^2$-fold *metal-enrichment* compared with solar abundances, i.e. look like the glowing ashes from exhaustive nuclear burning. Next, there are the problems of (ii) the *missing mass* at the centers of nearby galaxies, some $10^{6.5\pm1.5}$M$_\odot$ instead of the expected $10^{10.5\pm1.5}$M$_\odot$ from the past quasar fires; (iii) the *strengths* of their *winds*, inferred from the high outflow rates through the BLR which conform with $\dot{M}_{in} = \dot{M}_{out}$; (iv) their high γ-*ray compactness*, which would degrade the relativistic pair plasma when trying to escape from near the black hole's horizon in order to blow the jets (of the radio-loud sub-population); (v) the *hard spectra*, occasionally peaking above TeV energies: the heat radiation of a swallowing black hole is expected to culminate near keV$(M_\odot/M)^{1/4}$; (vi) the *inverted evolution* of the QSO phenomenon, whose Eddington power would grow with time whilst its observed power decreases rapidly; and (vii) the *nuclear-burning hurdle* which should at all times have reduced the compact core masses of galaxies via nuclear detonations, long before the critical density for collapse had been reached. Besides, (ix) the *universality* of the *jet phenomenon* suggests a universal engine which we know is a fast rotating magnet in the cases of newly forming stars, binary neutron stars, and forming binary white dwarfs [Kundt, 1996, 2000].

For the nine listed reasons, I share the doubts of a few other people, among them (the late) Viktor Ambartsumyan and Hoyle et al. [2000], in the widely accepted black-hole paradigm. The QSO phenomenon may rather be a straightforward consequence of the spiralling-in of matter through *galactic disks* whose density approaches stellar values near their centers, giving rise to almost relativistic Kepler velocities on the innermost *Solar-System scales*, to strong magnetization and magnetic reconnections, and to nuclear burning in the disk's midplane. Magnetic reconnections create the pair plasma which is responsible for the jet phenomenon and for the hard, non-thermal spectra, whereas nuclear power is used to re-eject matter at higher than SN velocities through the BLR. Active galactic nuclei may owe their extreme properties to those of their central disks.

10. High-Energy Radiation

Chemical fires liberate $\lesssim$ eV per atom, producing local temperatures near 10^4 K. Nuclear fires liberate $\lesssim$ MeV per atom, corresponding to temperatures near 10^{10} K. But Earth is hit by ions of kinetic energies $\lesssim 10^{20.5}$ eV, and by photons of energies $\lesssim 10^{16}$ eV, neither of which can be the offspring of a thermal process. In the absence of the detection of a high-energy engine at work, the scientific community has considered multi-step acceleration processes at shock surfaces as boosters, called *in situ*, or *Fermi accelerations*, comparable to ping-pong balls bouncing back and forth between approaching rackets. When individual reflections are elastic, this mechanism leads to an exponential increase in a particle's kinetic energy. But the second law of thermodynamics suggests that the assumption of elastic reflections becomes unrealistic when the test-particle regime is left, increasingly so with increasing energy excess of the charges over background energies [Kundt, 1998b; Hoyle et al., 2000]. It is my conviction that the cosmic rays gain their extreme energies in single-step boosts, in the corotating *magnetospheres* of fast-spinning compact stars, and that relativistic pair plasma is created in *magnetic reconnections*, as observed on the Sun.

10.1 The Cosmic Rays

The *cosmic-ray spectrum* impacting on the heliosphere has been plotted in Fig. 1.3, $E^2 \dot{N}_E$ versus ion energy E. At Earth, its sub-relativistic branch oscillates up and down with the 11-yr solar cycle and has been corrected for solar-wind screening, but its extremely relativistic branch is time independent. Above 10^{19} eV, its incoming energy flux is down by over seven orders of magnitude compared with its peak (at several GeV), but this ratio is not typical for the sources: Cosmic rays are stored in the Galactic magnetic fields for some 10^7 yr up to ion energies beyond which their Larmor radii $R_\perp$ grow comparable with fieldline curvature radii,

$$R_\perp = \gamma m_0 c^2 \beta_\perp / eB = 2 \text{ kpc } \gamma_{10} \beta_\perp (m_0/m_p)/B_{-5.3} , \qquad (10.1)$$

which happens near 10^{19} eV for protons ($\gamma = 10^{10}$). At these and higher energies, cosmic rays traverse the Galactic Disk within some $10^{2.5}$ yr, i.e.

spend $10^{4.5}$-times less time in it than their low-energy cousins so that their relative energy share is some $10^{-2.5}$, a non-negligible fraction! The cosmic-ray engines must invest a significant percentage of their power at the highest energies. Why then worry about the rest of the spectrum, a likely spillover from above?

A lack of conspicuous-looking Galactic boosters has led people to consider the possibility of extragalactic ones even though there is an under-abundance of cosmic rays in the SMC, by at least a factor of five. Note that the 3 K photon background limits the source distances to $\lesssim 10^2$Mpc, because of $e^{\pm}$-pair formation above 10^{18}eV, and even more so because of photoproduction of pions above 10^{20}eV, see (3.6) and (3.19). We are then faced with the conundrum of the observed *isotropy* in their arrival directions (above 10^{20}eV), obtained from air-shower statistics, which favours a large number ($>10^3$) of nearby boosters (in view of their limited range). Moreover, some 20% of these highest-energy cosmic rays have already *repeated*, i.e. arrived from the same direction within $\lesssim 1°$. Finally, the strong under-abundance of hydrogen and helium in the cosmic rays requires selective boosting, a problem for in situ models.

Once we consider nearby point sources, there is a clear preference for neutron stars because their potential *boost energy* ΔW (holding for particles of charge e):

$$\Delta W = e \int (\boldsymbol{E} + \boldsymbol{\beta} \times \boldsymbol{B}) \cdot d\boldsymbol{x} \le e \; \beta_\perp \; B \; \Delta x = 10^{21} \text{eV} \; (\beta_\perp B)_{12} \; (\Delta x)_7 \tag{10.2}$$

exceeds that of all other celestial bodies for corotational velocities $v = c\beta \le c$ inside the speed-of-light cylinder, and for (pulsar) surface magnetic field strengths $B \lesssim 10^{13.8}$G. Admittedly, the numbers inserted above are at the high end of expected values. But one can imagine situations in which clumps of radially infalling matter (from an imperfect SN ejection) indent the corotating magnetosphere, get squeezed to white-dwarf densities, and strain the magnetosphere whilst cutting their way diamagnetically down to the surface. The indented magnetosphere subsequently snaps back like a relativistic *slingshot* and expels all the ionized vaporization products. Its field strengths are twice the unconfined values at maximal confinement, and its expected runways (at some 60° ahead of the radial direction) are $\lesssim$ the speed-of-light radius $= 10^{7.5}$cm$/\Omega_3$. The thus-ejected cosmic rays are likely to escape through self-rammed, almost straight vacuum channels, hence will not have prohibiting curvature-radiation losses during the boost and escape. This scenario has similarities with magnetic accelerations discussed in Sect. 5.3 and also with recently observed transient sunward plasma motions in the magnetotail of Earth; see [Kundt, 1997, 1998b] for more detail.

Note that (10.2) could be formulated without the second term: Lorentz forces do not perform work. It has been written in its Lorentz-covariant form in order to facilitate the transition from a corotating frame to a global inertial frame in which only the electric field accelerates the escaping charges. Equation (10.2) holds equally for multi-step electromagnetic accelerations for which intermediate accelerations of either sign cancel out, and the net gain is again given by this expression, usually with much smaller values of $\beta_\perp B$ and net Δx; *in situ* estimates assume implicitly that such cancellations can be ignored.

10.2 The γ-Ray Bursts

The second high-energy conundrum of present-day astrophysics are the γ-ray bursts which reach Earth steadily from all directions, some three per day, with fluctuating peak energies straddling MeV but with occasional delayed hard tails up to 10 GeV and beyond, see Fig. 10.1, and again with intriguing *isotropy* of arrival directions yet only a rather limited range of fluxes, corresponding to a *thin-shell distribution* of the sources (spread by little more than an octave in typical distances). Their high temperature is remarkable, manifest through their low X-ray to γ-ray ratio: energy sharing with ambient material, e.g., from the hard surface of a neutron star, would bring the temperature down into the X-ray range.

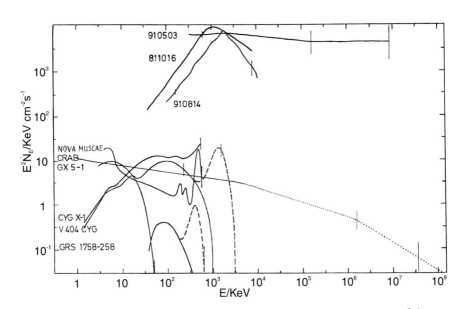

Fig. 10.1. Energy (or 'fluence') spectra of the Gamma-Ray Bursts, $\log(E^2 \dot{N}_E)$ vs $\log E$, seen in comparison with several spectra of accreting neutron-star (and BHC) sources as well as of the Crab nebula

The bursts have *luminosities*

$$L \lesssim 10^{38} \text{erg/s} \ (d/0.3 \text{ kpc})^2/\gamma^2 \tag{10.3}$$

if at distances $d = 0.3$ kpc and emitted isotropically ($\gamma = 1$), reminiscent of the Eddington limit of a solar mass. Beaming at a bulk Lorentz factor $\gamma \gg 1$ would reduce the involved power, but by what mechanism? If by jets, the latter would have to be hard, unlike those of the bipolar flows. Burst *durations* vary between $10^{-1.7}$s and 10^3s, with a two-humped histogram (like the spin periods of pulsars), but such that the time-integrated luminosity, or emitted *energy*, corresponds to an effective burst duration of 1s:

$$\int L dt \lesssim 10^{38} \text{erg} \ (d/0.3 \text{ kpc})^2/\gamma^2 \ . \tag{10.4}$$

Each two bursts have different, mostly *composite lightcurves*, reminiscent of neutron-star accretion of tidally torn clumps, like Jupiter's accretion of the 11 big pieces of comet Shoemaker-Levy 9 in July 1994 (of which more than 24 pieces were recorded). Sub-bursts have the appearance of FREDs (=fast rise, exponential decay), with temporal fine structure down to $10^{-3.7}$s, again reminiscent of (the small size of) neutron stars. Burst *spectra* are like those of cooling sparks, i.e. power-law superpositions of blackbody spectra with temporally decreasing temperatures during sub-bursts.

During the 1980s, the bursts were interpreted as being emitted by nearby Galactic neutron stars, during spasmodic accretion. The interpretation was backed up by reports of spectral lines at energies $\lesssim 0.5$ MeV, suggestive of redshifted pair annhilation, and at energies between 0.02 MeV and 0.1 MeV, suggestive of electron-cyclotron radiation in strong magnetic fields ($\gtrsim 10^{12}$G). It was independently backed up by the subclass of the (by now $\geq$ five) *soft repeaters*, a class whose members have made their first appearances via the strongest ever flashes, indistinguishable from all the other bursts, but have since repeated many times a year, in the form of somewhat softer and shorter, some 10^3 times fainter repetitions. They look like neutron-star binaries by their spectra at radio, optical, and X-ray frequencies, by their spin periods (between 5 s and 8 s), jet formation, and by being surrounded by synchrotron nebulae. Are they not just the nearest among all the bursters?

This conjecture was torpedoed by BATSE's improved statistics which found an isotropic, thin-shell distribution in the sky, unlike being hosted by the Milky Way. The sources 'must' thus be either very near, not much farther than 0.3 kpc, or else be *extremely far*, near the edge of the observable Universe. In the second case, their powers are raised by factors of order 10^{16} (through the corrected distances), way in excess of the brightest QSOs, and emitted at temperatures of order 10^{10}K (!), driving some people to look for alternative interpretations. For instance, if the sources are located in the Galactic Disk but radiate preferentially towards high Galactic latitudes, their appearance to terrestrial observers would be almost isotropic: we are not in the beam of the more distant sources.

Ad hoc as this assumption may sound, it would be realized by a population of *neutron stars* that oscillate through the *Galactic Disk* and accrete from interstellar clouds: The assembled, low-mass disks would be oriented preferentially at right angles to the Galactic Disk, and ricochetting chunks of accreting matter would radiate in a mildly beamed fashion, preferentially parallel to their tangent directions, at almost relativistic orbital velocities, and at low heights above the stars' surfaces [Kundt, 1998a]. The earlier interpretation via nearby Galactic neutron stars would be saved, and their otherwise undetectability explained at the same time. Moreover, the energetics, time scales, lightcurve morphologies, and spectra would all be consistent with neutron-star physics, rather than asking for a new, energetically detached class of phenomena. Statistics can lead in the wrong direction.

Still, it would be unfair not to mention the many laborious source 'identifications' with host galaxies, and the many burst *afterglows* with measured large absorption *redshifts*, monitored for days through months after the bursts. How many of the host galaxies are chance projections? This suspicion does not extend to the broadband afterglows seen in the long-duration subclass. But the afterglow lines are too narrow for galactic disks; and neutron-star winds are known to be relativistic – see SS 433 – hence qualify as local absorbers with redshifted transverse-Doppler shifts. Most of the energy of a spasmodic crash should be channeled into a transient relativistic hadronic wind composed of both surface and impacting material, centrifugally driven, which is overtaken by later segments of the burst if the latter is of long enough duration, $\gtrsim \Omega_{n*}^{-1} \approx 1$ s, thus giving rise to a broadband afterglow with a structured lightcurve, reminiscent of the light echo from a (super-)nova. I judge the energetics problem to be more serious than the statistical one.

This section should not end without a mentioning of the small subset of terrestrial γ-ray bursts, some 1% to 2% of all, which have been received from mid-atmospheric lightning, i.e. from discharges between the tops of large thunderclouds and the ionosphere. Note that they come from within the *terrestrial capacitor* – between the Earth's surface and the ionosphere – which is permanently charged to (0.6 ± 0.1) MV, the ionosphere being the positive pole. The bursts' spectra have harder slopes than the extraterrestrial ones, by spectral index one, but turn over below 1 MeV, and their durations are shorter, $\lesssim 5$ ms (instead of $\gtrsim 50$ ms). They are apparently generated on length scales of order 10 km, near the upward-propagating, inclined blue jets. γ-ray bursts do not necessarily require exotic sources.

11. Bipolar Flows

Bipolar flows – or *jet sources* – are abundant in the Universe. They first made their appearance in the early 1960s, as (1) extended, *extragalactic radio sources*, of sizes between $\gtrsim 10^2$pc and $\lesssim$ Mpc, powered by quasars. Miniature copies of them were discovered less than 20 years later, mostly at optical and infrared frequencies, powered by (2) young binary neutron stars (like SS 433) and BHCs, (3) very young ordinary stars (pre-T-Tauri stars, YSOs), and probably even (4) forming binary white dwarfs, at the centers of planetary nebulae. The stellar-mass ones also tend to be called *micro quasars*. The jet family therefore consists of 4 distinct classes; see Fig. 11.1, Plate 10, and the cover picture.

Whereas jet sources are detected at quite different frequencies, both line and continuum, they all share the following properties: They have (i) elongated *morphologies*, from unresolved *cores* through *hotspots* (knots, Herbig–Haro objects, FLIERs = fast low-ionization emission-line regions) to outermost *heads*, with jet *opening angles* θ of order 10^{-2}. The narrow jets never *branch*, and are contained in distinctly wider *lobes*, or *cocoons*, of axis ratios typically 1:5. The *cores* of the jet sources tend to radiate very (ii) *broadband*, and (iii) *variable*, often with spectra extending from low radio frequencies all the way up to $\gtrsim 10$ TeV photon energies in the extragalactic sources, and with (iv) *core/lobe* power ratios $L_{core}/L_{lobe} = 10^{2\pm2}$ corresponding to a mean jet-formation efficiency of 1%. The jets and hotspots often show (v) *sidedness*, i.e. they are much fainter, or often invisible on the receding side, even though the environment can be transparent. The knotty jets often (vi) grow, or expand *superluminally* (in projection).

11.1 How to Explain the Jet Sources?

When jet sources were more and more resolved and mapped, discoverers were impressed by their elongated morphologies, by the narrowness of the jets, and by the compactness and velocity ranges of the knots. How are the jets and lobes blown? Why are they stable? Two extreme possibilities offered themselves, depending on the mass-density ratio Ω of the jet fluid and the ambient medium: $\Omega \gg 1$, or $\Omega \ll 1$? The first possibility, a *hard beam*, is realised by firing bullets, or by a *lawn sprinkler*: When successive generations of bullets,

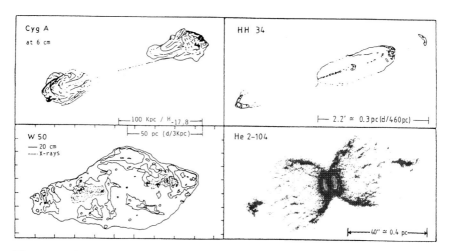

Fig. 11.1. Sketch of representatives of the Bipolar-Flow family, one for each class, together with their names, their sizes, and spectral information. Cyg A is mapped at radio frequencies, HH 34 and He 2-104 (Southern Crab) at optical emission lines, and SS 433 (inside W 50) in addition at X-rays. The literature contains $\lesssim 10^3$ well-mapped radio galaxies and quasars (AGN sources), $\gtrsim 10^2$ jets from Young Stellar Objects (YSOs), and $\gtrsim 10^{1.5}$ jets from binary neutron stars (and BHCs) and forming binary white dwarfs

or water drops are fired in varying directions, all observed beam shapes can be simulated, and an absence of branching is implied. The second possibility, a *soft beam*, is realized by blowing some light medium into a heavier one, as by a hot *hair drier*: now the shape of the beam is almost independent of a varying injection direction, like smoke rising from a chimney whose shape is controlled by the ambient winds.

Hard beams are much easier to model than soft beams. They seemed to be indicated by sources with regular shapes and by the neutron-star binary SS 433 with the periodically varying Doppler shifts of its blue- and redshifted recombination lines. Yet the implied energetics exceeded other estimates by orders of magnitude, and no plausible mechanism was offered for the firing of bullets, or for the observed multiple refocusing of the beams along their way out. Moreover, hard-beam models are forced to assume repeated *in situ (re-) acceleration* of electrons to the Lorentz factors inferred from the spectra, reaching 10^6 and more in strong extragalactic sources.

No consensus has yet been reached in the literature on the mechanism of jet formation, their substance, speed, energising, and focusing, with alternative interpretations allowing for (i) a {hadronic, leptonic} jet substance, (ii) electron Lorentz factors γ ranging from $\ll 1$ through $\gtrsim 1$ to $\gg 1$, (iii) {sub, super}-sonic propagation, (iv) with or without mass entrainment from their surroundings; see Begelman et al. [1984], Bahcall and Ostriker [1997], Kundt

[1996, 1998b]. My own preference is distinctly for the last entries, in particular for a soft beam that avoids the problem of electron (re-) acceleration which I consider in conflict with the second law of thermodynamics: Energization takes place exclusively in the central engine, in the form of localized magnetic reconnections – like in solar flares – which generate a relativistic electron–positron pair plasma of very high Lorentz factors. Such *pair plasma* is practically *weightless*, being

$$\rho_e/\rho_H = 6kT/m_p c^2 = 10^{-8.3} T_4 \tag{11.1}$$

times lighter than hydrogen of temperature T at the same pressure. (In the NR case, densities scale inversely as temperatures for fixed pressure). And it is *immiscible* with other plasmas because of frozen-in electromagnetic fields which it acquires during formation and during subsequent channel-wall interactions, and convects via minor velocity differences of electrons and positrons, $| \beta_+ - \beta_- | \approx 10^{-10}$ at equipartition of (particle and field) pressures.

Even for Lorentz factors as high as 10^6, such pair plasma can propagate almost loss-free through evacuated channels of lengths $\lesssim$ Mpc, via an ordered $\boldsymbol{E} \times \boldsymbol{B}$-drift, the only losses being inverse-Compton on the background radiations. No in situ acceleration is required: according to (3.4), inverse-Compton losses on the 3 K background have an e-folding length of

$$ct_{deg} \overset{(3.4)}{=} 3m_e c^2/4\sigma_T u_{3K} \gamma = \text{Mpc}/\gamma_6. \tag{11.2}$$

The synchrotron spectra of the hotspots and heads of the (extended) extragalactic radio sources imply power-law distributions of the *electron Lorentz factors* γ of exponent $g \approx 2.5$, see (3.31), between $\gtrsim 10^{2.5}$ and $\lesssim 10^7$ in the not-too-large sources, so that the high-energy tail dominates in the emitted power whilst the low-energy tail dominates in the energy density, or pressure.

The *central engines* of the four classes of bipolar flows listed above may all be *rotating magnets.* This statement is clear for the *young stellar objects* (YSOs) for which statistics says that every newborn star passes through this stage, including our Sun in her past. Forming at the center of its accretion disk, a young star is expected to have a (minimal) spin period of $P \approx \pi/\sqrt{G\rho}$ ≈ 3.6 h and to be strongly magnetized; its corona should therefore be vastly more active than that of the present Sun which has been spun down to $P = 27.3$ d, (the Sun's present *Carrington period*). Likewise, young *neutron stars* and even forming *white dwarfs* can have large magnetic moments and high spin rates and thereby create high voltages in their corotating *magnetospheres* (in interaction with the inner edges of their accretion disks), high enough for abundant pair creation; see problem 3.1.3, and Plates 3 and 4. There thus remains the class of *active galactic nuclei* (AGN) which are commonly thought to harbour supermassive black holes but may, instead, simply be the high-density, nuclear-burning centers of galactic disks; their (differentially) corotating *coronae* are plausible candidates for abundant pair formation [Kundt, 2000]. Two spectra are shown in Fig. 11.2; see also Plate 10.

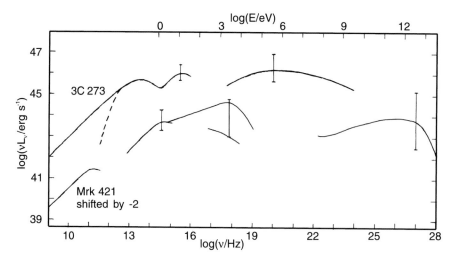

Fig. 11.2. Spectra of 3C 273 and Mrk 421 – $\log(\nu L_\nu)$ vs $\log \nu$ – two AGN of the (unscreened?) Blazar subfamily. 3C 273 is thought to have its synchrotron peak at visible frequencies and its inverse-Compton peak at GeV photon energies (red type) whereas Mrk 421 has the corresponding peaks at X-rays and TeV energies (blue type). Due to their rapid variability, simultaneous spectra are difficult to obtain; but it is clear that their spectra range among the broadest in the Universe, involving relativistic electrons of Lorentz factors $\lesssim 10^6$ and higher. The red quasars (3C 273) tend to be brighter than the blue ones (Mrk 421); note, however, that the latter has been shifted down in luminosity by a factor of 10^{-2}, to avoid overlap

It is therefore plausible that abundant *pair plasma* is created by all four classes of central engines, during their active epochs, with typical Lorentz factors of $10^{3\pm2}$ on escape. It is the most robust mechanism ever proposed. The pair plasma is created at the centers of local potential wells, at extreme pressures – many orders of magnitude above those of their surroundings – and tries to escape along the route of easiest escape which is the route of largest density gradient, along the spin axis, at right angles to the feeding accretion disk. In so doing, pressure confinement by the ambient substratum gives rise to the formation of a *deLaval nozzle,* first proposed by Blandford and Rees in 1974, beyond which the pair plasma escapes supersonically, i.e. faster than at $2c/3$, with *bulk Lorentz factors* comparable to the thermal Lorentz factors before the nozzle, in the relativistically hot central cavity.

Such forming jets have meanwhile been resolved by the HST in the stellar case, at initial diameters of $\gtrsim 10^{14}$cm (HH 30 and 34: [Burrows et al., 1996: Astrophys. J. *473*, 437]), and by the VLBA in the case of M 87, at an initial diameter of $\gtrsim 10^{16}$cm [Junor et al., 2000: Nature *401*, 891–892]. They plough two *antipodal channels* away from their origin as long as their ram pressure L/Ac in the terminal hotspot of their head (of aspect area A) exceeds the encountered (static or) ram pressure ρv_h^2, with a *head speed* v_h given by

$$v_h = \sqrt{L/Ac\rho} = 10^{-2}c \, (L/An)^{1/2}_{3.5} \qquad (11.3)$$

which can vary between $\lesssim 10^2$km/s and $\lesssim 10^{-1}c$, depending on the power of, and closeness to the central engine, and on the mass density $\rho = mn$ of the CSM, or IGM. For $3C$ 273 and very few others, head speeds close to c have even been considered because there is no visible trace of their counter jet, at dynamical contrasts of $\lesssim 10^{-4.5}$. Could it be bent, and hidden in projection by the approaching jet? In all other cases, the observed knot and head velocities are much smaller than c; they are the velocities of the pushed ambient medium, not of the jet substance. The latter is often invisible when it performs an ordered, radiation-free $\boldsymbol{E} \times \boldsymbol{B}$-drift, with toroidal magnetic fields plus often some longitudinal contribution, and radial electric Hall fields (w.r.t. cylindrical symmetry), see Fig. 11.3b. One thus understands why (even) relativistic beams need not radiate unless they encounter resistance – apart from unavoidable inverse-Compton losses on the radiation background – and why we often see knot velocities comparable to stellar-wind velocities even though the jet fluid moves luminally; see Fig. 11.3a.

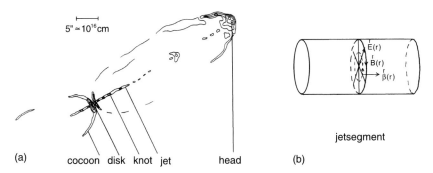

(a)   cocoon disk knot jet   head   (b)

Fig. 11.3. (a) Qualitative sketch of the geometrical building blocks of a Bipolar Flow, involving a hot core at the center of a feeding disk, the latter at right angles to the central beam, a chain of knots (hotspots, HHs, FLIERs) on at least one side, occasional (quasi-terminal) heads on one or both sides, and an elongated cocoon (lobe) which is thought to expand supersonically w.r.t. the CGM for type-A sources (see text). Observed scales range from $\lesssim 10^{14}$cm to $\gtrsim 10^{19}$cm in the stellar cases. (b) An ordered $\boldsymbol{E} \times \boldsymbol{B}$-Drift parallel to the beam, as a suggested – though controversial – realization of a bipolar flow, whereby the bulk velocity $c\beta$ may decrease radially from the axis to the periphery

At the same time, pair-plasma jets are understood to show *sidedness*, via beamed relativistic radiation along their tangent vectors which can be scattered – by filamentary obstacles – into all forward directions $\lesssim 90°$, and to show apparent *superluminal motion* at speeds $c\beta_\perp$, as a consequence of relativistic kinematics:

$$\beta_\perp = \frac{\beta \sin\theta}{1 - \beta \cos\theta} \leq \left\{\begin{matrix} \beta\,\gamma \\ ctg\theta \end{matrix}\right\}. \tag{11.4}$$

I.e. superluminal speeds are maximal at $\beta = \cos\theta$ – found e.g. by maximization of $1 + \beta_\perp^2$ w.r.t. $\cos\theta$ – for which $\beta_\perp = ctg\theta$, and $\gamma = 1/\sin\theta$. Equivalently, the inequality $\gamma \geq \sqrt{1 + \beta_\perp^2}$ holds strictly, and observed values $\beta_\perp \lesssim 30$ imply (occasional) bulk Lorentz factors $\gamma > 30$. The distribution of observed $\beta_\perp$ asks for yet larger values of γ, in excess of 10^2, in particular because inclination angles θ smaller than $1°$ are unrealistic for the wiggly jets.

Pair-plasma jets also provide the observed *hardspectra* wherever the beams are tapped, at reasonable overall power, out to the terminal hotspots where their motion is randomised, from supersonic to subsonic; no post-acceleration is required. The beam particles act like the electrons hitting a TV screen. They then blow the *cocoons*, still with considerable Lorentz factors – and at supersonic speeds w.r.t. the ambient medium – but fade due to adiabatic expansion and due to radiation losses.

This soft-beam interpretation is not commonplace and should be supported by a few more facts. For instance, two (young) stellar *radio-triple* sources are known, indistinguishable from the extragalactic ones except for their cores; and *one-sided radio jets* have been detected in their cores and in that of HH 111. Axial (optical) hotspots have been mapped in the core of the planetary nebula M 1-92 [Trammell and Goodrich, 1996] and of the PNe He2-90 and CRL 618, and *superluminal expansions* in the radio jets of at least five Galactic BHCs which I interpret as neutron stars inside of massive disks [Kundt, 1998a]. Protostellar jets (in Orion) can be destroyed by the radiation from nearby bright stars [Reipurth et al., 1998], via inverse-Compton losses, and similarly the counter jet in the Cyg X-1 system. Molecular-line emission is mapped in stellar sources with velocity dispersions of $\lesssim 10^{2.7}$km/s whereas a hard impact, at $\gtrsim 10^{1.3}$km/s, should suffice to decompose the molecules; an *ultrasoft acceleration* is indicated. Finally, the Milky Way is an emitter of the 511keV γ-ray line of pair annihilation, 10^{43}s^{-1}, known for a number of years without having found any other plausible explanation (at this high rate): Such annihilations are expected from all the Galactic jet sources, millions of years after their active epochs, when the relativistic charges have diffused through large distances and been slowed down to non-relativistic speeds. Relativistic beams are not easy to detect, but are often required by the remarkable details of the observations as well as by the overall dynamics.

Bipolar flows show various *morphologies*, often quite different from the simplest, straight one sketched in Fig. 11.3. For a soft beam, it is clear that during its early stages, transverse motions of the ambient medium can easily bend its channel and cocoon because of its low inertia, and because of the comparatively slow advance of its termination shock (head). On galactic scales, the beam can thus acquire S-shapewhilst on intergalactic scales, the ambient 'wind' can bend it into the shape of a letter U.Later generations of jet particles have a tendency to *straighten* the channel, by rubbing against the bending wall; this causes a deformation towards the shape of an I,with the hardest spectra found in the straightest segments.

Apart from straightening with age, there is a *dichotomy* among the extragalactic source morphologies concerning their outer parts: is the jet-lobe morphology conserved all the way out to the head, with a sharp front edge (type A), or does the jet with its knots gradually disappear and give way to a dispersing, blunt cocoon (type B)?This dichotomy, due to Jean Eilek [1999], slightly modifies the earlier, purely phenomenological classification by Fanaroff and Riley [1974] as of class I and class II – whereby class II is a subcase of type A – and may well correspond to the advance speed of the head being *super-*, or *subsonic* w.r.t. the penetrated medium: in the latter case, the jet substance is no longer reflected at its termination bowshock, on transition from super- to subsonic propagation, but can expand more or less isotropically beyond the shock surface, thus blowing a smooth, blunt, elongated cocoon.

12. Image Distortions

Our *electromagnetic maps* of the sky, at whatever frequency, are distorted from their intrinsic geometry by the time-varying inhomogeneity of the intervening medium, via diffraction, refraction, and scattering: compact sources *scintillate*. Moreover, General Relativity implies that additional, non-dispersive distortions occur due to the inhomogeneities in the cosmic-mass distribution, called *gravitational lensing* [Schneider et al., 1992].

12.1 Scintillations

Because the Universe is neither empty nor filled homogeneously with matter, electromagnetic signals suffer frequency-dependent distortions during propagation; i.e. our records of distant objects are perturbed. Such scintillations are caused by the Earth's *atmosphere* and *ionosphere*, by the interplanetary medium (*IPM*), by the interstellar medium (*ISM*), and by the 'weather' immediately in front of a source. Weak inhomogeneities impose *diffractive* distortions, strong inhomogeneities impose *refractive* distortions and/or *scattering* halos. Such distortions can be strong for point-like sources, and cancel increasingly with a growing size of the source. For instance, distant street lights scintillate if bulb-sized, but are steady if more extended. Stars tend to scintillate whereas our fellow planets do not. Scintillations can therefore be used to estimate the size of a source and the irregularities and velocities of the intervening media.

Scintillations result from inhomogeneities in the *refractive index* $n = \sqrt{\epsilon\mu}$ – n to be distinguished from electron-number density n_e – where the dielectric constant ϵ is determined by the resonances in the medium according to (3.14) and (3.16) for $\hat{a} \ll 1$, viz.

$$-\delta n = 1 - \sqrt{\epsilon\mu} \approx \omega_{pl}^2/2\omega^2 = r_e\lambda^2\Delta n_e/2\pi = 10^{-13.8}\,(\Delta n_e)_{-3.5}/\nu_9^2 \quad (12.1)$$

with $\Delta n_e := \sqrt{<\delta n_e>^2}$ = mean-squared fluctuation of electron-number density, and with the numerical value holding for typical interstellar scintillations at frequency $\nu := c/\lambda = $ GHz. We shall now evaluate approximate expressions for the expected angular and temporal image distortions.

When an electromagnetic wave of phase factor $\exp\{i\Phi\} = \exp\{i(\boldsymbol{k}\cdot\boldsymbol{x}-\omega t)\}$ crosses a plasma blob of size a and refractive-index contrast δn, it picks up a phase delay $\delta\Phi$ of order $\delta(\boldsymbol{k}\cdot\boldsymbol{x}) = \delta k\,a = k\,a\,\delta n/n$ because of $\boldsymbol{n} = c\boldsymbol{k}/\omega$, and is thereby deflected through a small angle $\theta_{sc} = (\lambda/a)(\delta\Phi/2\pi) \approx \delta n$ (for $n \approx 1$). When the scintillation screen has a thickness s, a light ray traversing it will be deflected some s/a times so that its angle increases randomly by the factor $\sqrt{s/a}$, yielding

$$\theta_{sc} \approx \mid \delta n \mid \sqrt{s/a} \approx (r_e\lambda^2\Delta n_e/2\pi)(s/\lambda)^{1/4} = 10^{-8.8}\,(\Delta n_e)_{-3.5}\,s_{21.5}^{1/4}\,/\,\nu_9^{7/4}\,. \tag{12.2}$$

Here the fluctuating blob size a has been approximated by the size of the first Fresnel zone, $a \approx \sqrt{\lambda s}$, and s has been inserted in units of a kpc, the typical interstellar distance (to a pulsar). As a result, the random *scattering angle* θ_{sc} takes values of $\lesssim$ marcsec at GHz frequencies. Sources smaller in angular extent than θ_{sc} cannot be resolved in a map.

Scattering through an angle causes two delays of a signal: a detour delay, and a propagation delay due to a transiently smaller group velocity. For a source distance d and a thin scattering screen half-way between source and observer, the *detour delay* $\Delta_\theta t$ follows after Pythagoras as $\Delta s/c = 2(d/2c)$ $(\sqrt{1+\theta_{sc}^2/4} - 1) \approx (d/c)(\theta_{sc}^2/8)$. The *propagation delay* is the random sum of s/a successive delays of order $\delta(a/v_{gr})$, i.e. $\Delta_{gr}t = \sqrt{s/a}\,a\,\delta(1/v_{gr}) \approx \sqrt{as}\,\omega_{pl}^2/2c\omega^2$ for $\epsilon \approx 1 - \omega_{pl}^2/\omega^2$ because of $ck = n\omega \approx \sqrt{\omega^2 - \omega_{pl}^2}$, whence $dk/d\omega \approx \omega/c^2k$, $v_{gr} := d\omega/dk \approx c^2/v_{ph}$,

$$\beta_{gr} \approx 1/\beta_{ph} = n = \sqrt{\epsilon} \approx 1 - \omega_{pl}^2/2\omega^2\,, \tag{12.3}$$

and $\delta v_{gr}/v_{gr} \approx \delta\epsilon/2\epsilon \approx \delta\epsilon/2 \approx -\omega_{pl}^2/2\omega^2$. Adding the two delays together, we arrive at a total *pulse broadening* Δt of

$$\Delta t = (d/c)[\theta_{sc}^2/8 + (\sqrt{as}/d)(\omega_{pl}^2/2\omega^2)]$$
$$= 10^{-7.5}\sec \nu_9^{-4}d_{21.5}(s/a)_{10}(\Delta n_e)_{-3.5}^2 \tag{12.4}$$
$$\cdot\,[1 + \nu_9^2(a/d)_{-10}^{3/2}(d/s)^{1/2}/2(\Delta n_e)_{-3.5}]$$

with $s \leq d$ whose individual terms scale with frequency and distance as $\nu^{-\alpha}d^\beta$ with $(\alpha,\beta) = \{(7/2,3/2),(9/4,3/4)\}$ for $a \approx \sqrt{\lambda s}$, i.e. dominate at {low, high} frequencies. They coincide at the transition frequency $\nu_9 = [2(d/a)_{10}^{3/2}$ $(s/d)^{1/2}\,(\Delta n_e)_{-3.5}]^{1/2}$ whose value was found to equal 0.15 for the Vela pulsar in 1975, a reasonable agreement. The pulse broadening time scale Δt can also be measured as the inverse of the *decorrelation bandwidth* $\Delta\omega := 1/\Delta t$, viz. the angular-frequency bandwidth beyond which scintillations interfere destructively, i.e. wash out.

How extended must a source be in order to stop scintillating? A source of size l radiating at wavelength λ has its first diffractive minimum at an angle of order λ/l. Consequently, its own diffractive pattern will destructively interfere with that of an inhomogeneous medium unless λ/l is smaller than θ_{sc}. From (12.2) we thus get for the *critical size* l_{crit} of a scintillating source:

$$l_{crit} = (2\pi/r_e\lambda\Delta n_e)(\lambda/s)^{1/4} = 10^{10.3}\text{cm}\; \nu_9^{3/4}/s_{21.5}^{1/4}(\Delta n)_{-3.5}\;. \qquad (12.5)$$

For pulsars, scintillation measurements can thus determine the size l_{crit} of their emission region: Is it as small as polar-cap dipole models predict, some 10^{-2} of the speed-of-light cylinder's radius $c/\Omega = 10^9\text{cm}/\Omega_{1.5}$, or is it distinctly larger, corresponding to the gyro resonance (which is traversed further downstream), or is it broadened by scattering near the speed-of-light cylinder and hence comparable with its radius? There have been a few reports of refractive scintillation which favour large emission sizes.

Even for a static screen, a scintillation pattern *varies* with time when the screen moves transversely to the line of sight at speed $v_\perp$. Such patterns have been observed rushing across the Earth at various velocities, resulting from the Earth's motion around the Sun (at 30 km/s), the Sun's motion around the center of the Galaxy (at 220 km/s), the solar wind escaping from the Sun (at between $10^{2.3}$km/s and $10^{3.3}$km/s), inhomogeneities in the ISM (at $\lesssim 30$ km/s), and the source itself. The variability time scale $t_{scint} := l_{crit}/v_\perp$ is called the *scintillation time*; for different situations, it takes values between less than a second (Earth's atmosphere, in the visible) and hours (interstellar radio).

Analogously to delayed arrivals, the bandwidth $\Delta\omega := 1/t_{scint}$ associated with the scintillation time t_{scint} is called *scintillation bandwidth*; it is often shorter than mHz. Further, one talks of *strong scintillations* when the involved phase fluctuations are $\gg 1$, a situation for which more than one cone of scattered radiation reaches the observer at any time, or $\theta_{sc} > 2a/d$. From (12.2) with $s = d$, this inequality cannot be satisfied above the *critical frequency*

$$\nu_{crit} = c(r_e\Delta n_e/4\pi)^{1/2}(d/a)^{3/4} = 10^{9.5}\text{Hz}\;(\Delta n_e)_{-3.5}(d/a)_{10}^{3/4}\;, \qquad (12.6)$$

above which scintillations become weak. For weak scintillations, the *modulation index* $m := \sqrt{<(\Delta I)^2>/<I^2>}$ takes values $\ll 1$.

Scintillation theory has progressed a long way beyond the elementary estimates presented in this section. More elaborate studies treat scintillations as stochastic processes and take their departure from the (auto-) *correlation function* $B_n(\boldsymbol{x}_1, \boldsymbol{x}_2) := <\delta n(\boldsymbol{x}_1)\delta n(\boldsymbol{x}_2)>$ of the refractice index n, with $\delta n := n - <n>$, from the *spectral function* $\tilde{B}_n(\boldsymbol{k}; \boldsymbol{x}) :=$ Fourier transform of $B_n(\boldsymbol{x}_1 - \boldsymbol{x}_2, (\boldsymbol{x}_1 + \boldsymbol{x}_2)/2)$ w.r.t. $\boldsymbol{x}_1 - \boldsymbol{x}_2$, and from the *structure function* $D_n(\boldsymbol{x}_1, \boldsymbol{x}_2) := <[n(\boldsymbol{x}_1) - n(\boldsymbol{x}_2)]^2>$, all of which involve integrals over Bessel functions in the simplest applications. Even then, higher-order correlations are usually ignored which would be necessary for higher rigour. Still, much has already been learned from the elementary estimates presented above.

12.2 Gravitational Lensing

Electromagnetic signals are even distorted when propagating through perfect vacuum, by the presence of inhomogeneous mass distributions, because 4-d spacetime is curved by the presence of matter. In Einstein's theory of General Relativity, luminal signals are described by null geodesics which are everywhere tangent to the local light cones. On these, any localized mass concentration acts as a *gravitational lens*, bending their rays in 3-space as does the (collecting) foot of a wine glass. Like in flat spacetime, *Fermat's principle* (of hurried light) can be shown to hold, as well as a conservation of *surface brightness*. Magnification therefore implies an enhancement in brightness: strongly lensed objects look brighter than otherwise.

In principle, therefore, no map of the sky can be taken as a reliable documentation of what it contains. In practice, however, gravitational image distortions tend to be weak, and only one quasar in 500 is strongly magnified by foreground galaxies, by a factor of several. We presently know less than 30 strongly lensed quasars, each of which holds interesting information. An even smaller number of distant galaxies are distorted into *arcs*, by foreground clusters of galaxies. Among others, magnified objects allow a mass determination of their lens – independent of estimates via the virial theorem, or via the X-ray luminosity of its hot gas – and a determination of the Hubble parameter when sufficiently far. Lensing can be used to measure the 4-d geometry, and to trace *dark matter*, see Plate 12.

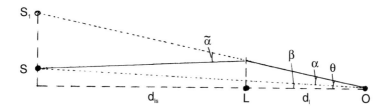

Fig. 12.1. Geometry of a point source S lensed by a foreground point mass L: its position (or impact) angle β in the sky is enlarged by the reduced deflection angle α, yielding the enlarged image S_1 at position angle θ. (There is also a mirror-inverted, demagnified image S_2 on the opposite side). The source distance d_s is the sum of lens distance d_l and lens-source distance: $d_s = d_l + d_{ls}$

In order to see this, note that a distant source, offset from a foreground lens by an (impact) angle β, is mapped at a slightly larger angle θ, i.e. magnified through a small *reduced deflection angle* $\alpha := \theta - \beta$, see Fig. 12.1. For a comparatively nearby lens, Einstein's light-deflection angle $\tilde{\alpha} = 4GM/c^2\theta d_l$ can be used to express $\alpha \approx \tilde{\alpha} d_{ls}/d_s$ through the involved distances d_j, and mass M of the lens, or alternatively through the lensing surface-mass density σ enclosed by the rays, as

$$\theta - \beta =: \alpha \approx \theta_E^2/\theta = (\sigma/\sigma_{crit})\theta ,\tag{12.7}$$

where the (angular) *Einstein radius* θ_E, and the *critical surface-mass density* σ_{crit} are given by

$$\theta_E = \sqrt{(4GM/c^2)(d_{ls}/d_l d_s)} \approx 2 \text{ arcsec } \sqrt{M_{(12)}/d_{28}}\tag{12.8}$$

and

$$\sigma_{crit} = (c^2/4\pi G)(d_s/d_l d_{ls}) \approx 0.8 \text{ gcm}^{-2}\, d_{28}^{-1} ,\tag{12.9}$$

and where $\sigma := M/\pi(\theta d_l)^2$. Here the indices l, s, and ls stand for lens, source, and lens-source respectively, $M_{(12)} := M/10^{12} M_\odot$, and $d_{ls}/d_s \approx 1$ has been assumed. θ_E is the angular radius at which a source appears as a ring in the sky – an *Einstein ring* – if it is located exactly behind a lens, i.e. for $\beta = 0$; half-a-dozen such rings are presently known, discovered at radio frequencies. The critical surface-mass density is useful in particular in application to (extended) galaxy clusters where it characterizes the smeared-out mass necessary for strong lensing; σ/σ_{crit} is called the *optical depth* (to lensing).

For point-like lenses, there are always *two images* whose apparent angles θ_j follow from (12.7) by solving the quadratic equation:

$$\theta_{1,2} = \frac{1}{2}\left(\beta \pm \sqrt{\beta^2 + 4\theta_E^2}\right) .\tag{12.10}$$

$\theta_{1,2}$ straddle the Einstein radius, with one of them negative, i.e. antipodal; its associated image has opposite *parity*. Their summed *brightness enhancements* (angular-area ratios) $\mu := |\,\mu_1\,| + |\,\mu_2\,|$ follow from $\mu = \theta d\theta/\beta d\beta = [1 - (\theta_E/\theta)^4]^{-1}$, where use has been made of $\beta/\theta = 1 - (\theta_E/\theta)^2$, as

$$\mu = \frac{1 + u^2/2}{u\sqrt{1 + u^2/4}} \geq 1 \text{ with } u := \beta/\theta_E \lesssim 1 .\tag{12.11}$$

μ scales roughly as $1/u$ for small '*impact parameter*' u, hence grows unlimited on approach of alignment ($\beta = 0$). Extended lenses can produce more than two images, as is familiar from cluster imaging, with alignment generalised to crossing a *caustic*. For them, brightness enhancement μ is measured by the inverse determinant of the Jacobian matrix $\partial\beta/\partial\theta$ of the lensed map, in which β, θ are 2-d vectors in the lens plane.

In 3-d language, lensing implies a detour plus a (Shapiro) delay during passage through the potential ψ of the lens, the two adding up to a total *delay* Δt, different for different images:

$$\Delta t = (1 + z_l)(d_l d_s / d_{ls} c)[(\alpha^2 / 2 + \psi(\theta)] \approx d_l \alpha^2 / 2c = 43 \, \text{day} (\alpha / \text{arcsec})^2 d_{28} \tag{12.12}$$

in which ψ is the (angular) potential of $\boldsymbol{\alpha}$, $\boldsymbol{\alpha} =: -\nabla \psi$, and must be modelled for a careful determination of the Hubble parameter. This is possible for lensing galaxy clusters with *giant luminous arcs* and multiple *arclets* each of which is a distorted image of a more distant galaxy, and offers redundancy.

So far we have described *strong* lensing, with multiple images. But in addition, light rays often pass close to stars, or dark objects in the halos of lensing galaxies, or even in our own Galaxy, or close to faint intervening galaxies, or even through gravitational waves, and give rise to *microlensing* on angular scales $\theta \sim \sqrt{M/d}$ between µarcsec and many arcsec, see (12.8). Microlensing can be detected by recording light curves which are distorted during close passage of a massive object through the line of sight. Such *lightcurve distortions* can be identified as lensed events by being *non-dispersive*, i.e. frequency independent. Statistical campaigns have revealed that between us and the Large Magellanic Cloud, the *optical depth* $\tau(\text{LMC}) := \sigma / \sigma_{crit}$ equals $10^{-7.2 \pm 0.2}$ – consistent with star-like lenses of average mass 0.5 $M_\odot$ – and that the corresponding depth towards the center of our Galaxy reads $\tau(\text{bulge}) = 10^{-6 \pm 0.3}$. A combination of strong lensing (by an intervening galaxy) and microlensing (by one of its stars) should yield lightcurve distortions which can occasionally resolve the BLR of a QSO, at $\theta d \approx 10^{17} \text{cm}$.

In contrast to what has been discussed so far, every map of a distant object is weakly lensed, by the omni-presence of inhomogeneities. *Weak lensing* can only be detected statistically. It can be used, e.g., for determining the cosmic mass-density fluctuation spectrum, through its shear distortion of galaxy shapes.

13. Special Sources

Research in modern astrophysics is strongly biased towards a handful of very *bright*, *broadband*, and often *variable* sources which are a lot easier to map and/or monitor than others because of their large fluxes. This chapter selects seven of them, all members of our Galaxy. Their correct interpretation is crucial for our understanding of the Universe.

13.1 The Crab Nebula and its Pulsar

The Crab Nebula is perhaps the most scenic and least typical among the ≥ 9 known Galactic SN shells formed in the past millennium (Sect. 13.3) – known as the *historical SNe* – or the several hundred known Galactic SN shells; see Plates 5 and 6 and Fig. 13.1. Its launch in 1054 was recorded by the Chinese and left the *Crab pulsar* at its center, of present period $P = 33.1$ ms. The Crab's pulsed and unpulsed spectrum range from $\lesssim 10$ MHz to at least 10 TeV, with past reports reaching up to 10^{16}eV which have not been confirmed recently. Its filamentary thermal shell is a non-trivial realization of Sect. 2.7.

The *thermal component* of the Crab nebula, of temperature $T \gtrsim 10^4$K, number density $n = 10^{3.5 \pm 0.5}$cm^{-3}, mass $M \gtrsim$ M$_\odot$, has been shown to perform a strict Hubble flow, $v(r) \sim r$, to be at a distance of 2 kpc for an assumed cigar shape (rather than discus shape), and to be post-accelerated by 8% at late times (for a launch in 1054). It consists of $\gtrsim 10^4$ magnetized filaments, of diameters from $10^{16.5}$cm down to $\lesssim 10^{14.5}$cm, volume-filling factor $f \lesssim 10^{-3}$. Its size and kinetic energy are atypically small for a SN, by factors of 5 and 5^2, respectively, and its post-acceleration is ill-understood. In my understanding, we see only the trailing filaments of the SN explosion, of velocities $\lesssim 1.8$ Mm/s, which are still inside the pair-plasma bubble, of 10^4-fold overpressure, $p \lesssim 10^{-8}$dyn/cm^2, but do not see the faster filaments, at several times larger radii, which move through the highly evacuated outer windzone of the Crab's blue progenitor [Kundt, 1990a; also: 1998: MEMSAIT *69*, 911-917].

This explanation of the Crab's seemingly atypical *dynamics* leaves its energetics in the typical range of a SN. Exceptional are the low *density n* and high expansion speed *v* of its CSM, according to

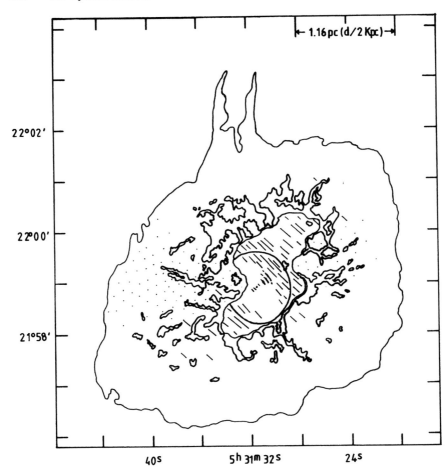

Fig. 13.1. Schematic map of the Crab Nebula, one of the youngest and best-studied SNRs, whose birth in 1054 was recorded in China. Its outermost contour line, at radio or [OIII]-line frequencies, traces the expanding boundary between the CSM and the nebula's pressurizing relativistic component – magnetized pair plasma? – which is also responsible for the strong linear polarization, and for the transrelativistically moving wisps, predominantly NW from the central pulsar (*short straight lines*). Linearly expanding, trailing SN ejecta are seen as line-emitting filaments, realizing a Hubble flow (*drawn wiggly*), and a fat expanding dust ring (*marked stippled*) is called 'dark bays'. The expansion of the Crab's shell of SN ejecta is atypically slow: $v(r) \lesssim 10^{3.26}$km/s; all the faster filaments are presently invisible, in the underdensity of the surrounding blue progenitor's evacuated windzone. See also Plates 5 and 6, and Astropys. J. *416*, 251–255 (1993), *456*, 225–233 (1996)

$$n = \dot{M}/4\pi r^2 vm = \begin{cases} 10^{-0.3}\text{cm}^{-3} \; \dot{M}_{(-5)} \; / \; r_{19}^2 \, v_6 \;\; , \text{red} \\ 10^{-3.3}\text{cm}^{-3} \; \dot{M}_{(-6)} \; / \; r_{19}^2 \, v_8 \;\; , \text{blue} \end{cases} \tag{13.1}$$

for {red, blue} supergiants respectively, $\dot{M}_{(a)} := \dot{M}/10^a (\text{M}_\odot/\text{yr})$; see (2.7). For the filaments' post-acceleration, an invisible piston is required that has transferred their huge radial excess momentum. No other agent has come to my knowledge than the wave bath of strong, multiply ($10^{1.5}$-times) reflected 30-Hz waves emitted by its central pulsar [Kundt, 1990a].

The fate of the Crab's 30-Hz *magnetic dipole waves* has been controversial throughout over 25 years. Do they post-accelerate the outgoing pair-plasma wind of the pulsar, from bulk Lorentz factors $\gamma \lesssim 10^3$ near the speed-of-light cylinder to $\gamma \gtrsim 10^{5.5}$, according to the Gunn–Ostriker mechanism of radial $\boldsymbol{E} \times \boldsymbol{B}$-pushing? I think so, for the following four reasons: (i) The hard spectrum of the Crab pulsar requires (many!) electrons of Lorentz factors $\gamma \lesssim 10^{8.8}$ for its emission via synchrotron and inverse-Compton processes, half of which are probably positrons (because of charge neutrality at low inertia, creation in vacuum, low circular polarization); their simultaneous boosting relies on $\boldsymbol{E} \times \boldsymbol{B}$-forces at and beyond the speed-of-light cylinder (which act equally on charges of either sign). (ii) The pulsar's measured spindown requires a torque comparable to the wave-emission recoil. (iii) The pressure in the nebula, sensed by its synchrotron emissivity, indicates a high energy density, corresponding to $\sqrt{< \boldsymbol{B}^2 >} = 10^{-3.3 \pm 0.3}$G, see (3.4). (iv) The post acceleration by 8% of the radially escaping filaments requires a strong radial momentum transfer, realisable by some 30 successive reflections of the 30-Hz waves from opposite sides of the nebula.

These 30-Hz *waves* are very *strong,* all the way out into the nebula (according to (iii), (iv) above):

$$f := eB/m_e c\Omega = 10^{11.7}/r_8 \tag{13.2}$$

holds for the strength parameter (3.20) of electrons near the speed-of-light cylinder, for a polar transverse dipole fieldstrength $B_\perp = 10^{12.6}$G falling off radially as r^{-3} out to $r = c/\Omega = 10^{8.2}$cm, and as r^{-1} beyond. They are expected to sweep all charges outward with them as long as the charges' inertia (Lorentz factor) stays inferior. Thereafter, a fair guess expects equipartition of 4-momenta among wave and charges – whose rigorous proof is still lacking – i.e. an energy-transfer efficiency of order 50%. Charges injected into a strong spherical vacuum wave have been shown to reach Lorentz factors of order $f^{2/3} \lesssim 10^8$; but here we deal with a *subluminal* ($E < B$) *strong plasma wave* which may well achieve the observed electron-injection spectrum into the Crab nebula at its inner edge, near $r = 10^{18}$cm:

$$N_\gamma d\gamma = \gamma^{-2.2} d\gamma \;\; \text{for } 10^{5.5} \lesssim \gamma \lesssim 10^{8.8} \; . \tag{13.3}$$

Note that a Lorentz factor of $10^{5.5}$ just guarantees that the electrons stay in phase within 10^{-2} wavelengths with the 30-Hz waves, on their way out into the nebula.

With this we have arrived at the third component of the Crab, its *pair-plasma wind*. According to (3.4), the *aging time* t_s of relativistic electrons at the inner shock due to synchrotron radiation is

$$t_s = \gamma/\dot{\gamma} = 4\pi m_e c/\sigma_T \gamma B^2 = 8 \text{ yr} / \gamma_6 B_{-3}^2 , \qquad (13.4)$$

i.e. short compared with the expansion time (of 10^3yr) for Lorentz factors $\gamma > 10^5$. The electron *injection rate* $\dot{N}_e$ can therefore be obtained from the Crab's luminosity $L = 10^{38.2}$erg/s via division by the average electron energy:

$$\dot{N}_e = L / <\gamma> m_e c^2 = 10^{38.5} \text{s}^{-1} . \qquad (13.5)$$

$\dot{N}_e$ equals $10^{4.1}$-times the Goldreich–Julian rate $\dot{N}_{GJ} = \mu\Omega^2/ec = 10^{34.4}\text{s}^{-1}$ for $\mu := BR^3 = 10^{31}$G cm^3 = the pulsar's magnetic dipole moment. This is a huge rate, because the Goldreich–Julian rate assumes an escape at the speed of light of the space-charge density which would be expected above the polar caps under steady-state conditions. One learns that pulsar winds exceed quasi-stationary winds by large factors, of order 10^4, hence must be driven in a pulsed mode, with ample particle production in vacuum [Kundt, 1998a]. Such particles should be dominantly electron–positron pairs. A similar estimate (of $\dot{N}_e$) can be obtained for ≥ 9 other pulsars from the mapped standoff radii of their bowshocks.

As already found in (11.1), pair plasma is as *weightless* as any relativistic fluid in pressure balance, some $10^{8.3}$-times lighter than warm hydrogen, hence it cannot account for the filaments' post-acceleration, not even with the (doubtful) hydraulic effect included. But it is thought to *fill the volume* of the Crab, and to pressurize it. The pair-plasma wind is thought to escape quasi-loss-free out to the inner edge of the Crab, at $r \approx$ lyr, where its supersonic flow is smoothly stalled into subsonic (relativistic) gyrations during which all the hard electrons degrade towards the radio regime of the spectrum. In this way, a significant fraction of the present $N_e = 10^{49.5}/B_{-3}$ radio-emitting relativistic electrons in the nebula have been continually supplied. The small deficit, $N_e - \int \dot{N}_e dt$, can be understood as due to a combination of an initial SN supply plus a somewhat stronger injection in the past.

Most of the Crab's emission comes from near the inner shock layer, strongest from an (inclined) equatorial torus, in the form of the almost luminally outward moving pattern of *wisps* which are brighter on the approaching NW side than opposite to it, due to a combination of relativistic beaming and an asymmetric radial pressure fall-off. Should the wisps be understood as a *laser* phenomenon, signalled by their high surface brightness? Does the northern *chimney* represent an outflow channel for the pressurized relativistic plasma? Are the *dark bays* the remnants of a massive gaseous *torus* around the Crab's progenitor? What is the origin of the single more redshifted filament (~ 2), with its 11 compact knots? Are the *anomalous pulses* of the Crab, between 4 GHz and 10 GHz, due to curvature radiation (at frequency $\nu = c\gamma^3/r$) during traversal of the inner one lightyear? Should the *jet-like X-ray* morphology of the Crab's inner part be considered a propagation anomaly near its spin axis? A wealth of detail problems wait to be answered. They can teach us the particulars of the interaction of a rotating magnet with its circumstellar medium, which bears strong similarities to a QSO engine.

There remains a fourth component to be mentioned: the *toroidal magnetic flux*, convected outward by the pair-plasma wind, as a frozen-in residue of the spin-parallel flux component. Its relative energy density has been judged to be low, of order 10^{-3}, from the (moderate) pressure in the nebula. The Crab pulsar is therefore thought to be an almost perpendicular rotator.

13.2 SS 433

SS 433 is the 433rd entry in the 1977 catalogue by Stephenson and Sanduleak of variable stars. It made headlines since the 1978 Texas Symposium on Relativistic Astrophysics held in Munich, mostly because of its two sets of periodically *blue- and redshifted hydrogen and helium recombination lines* in emission: how can matter move relativistically near a Galactic star? SS 433 is a *high-mass* radio and X-ray *binary* inside the SN shell W50, of orbital period $P = (13.0862 \pm 0.03)$d, shows a pair of precessing, transluminal jets on length scales between $\lesssim 10^{15}$cm and $10^{20.3}$cm (d/3kpc), of period $P_{prec} = (164.0 \pm 4)$d, also several low-order beat periods of the two including *nodding* (at $(2P^{-1} + P_{prec}^{-1})^{-1}$) plus correlated variabilities down to 5 min, and has meanwhile also revealed two sets of shifted, almost completely ionized X-ray emission lines from iron, nickel, magnesium, calcium, silicon, sulfur, argon, and neon. Most of its derived properties have been controversial for 20 years – such as its masses, distance, power, jet speed, mode of precession, and composition – despite several international conferences devoted solely to it; see [Kundt, 1996; also: 1999, MEMSAIT *70*, 1091–1103], and Fig. 13.2.

Among the presently less controversial properties of the system SS 433 are its composition of a B star orbited by a non-accreting (*ejecting*) neutron star, at the center of the 10^4yr-old W50 which has given birth to the neutron star, and whose (radio) ansae have been blown by the jets. Its *distance* has been determined as (3 ± 0.5)kpc by seven of the astronomical standard methods: 21cm absorption, interstellar (optical) absorption, soft X-ray cutoff, CO mapping, size of W50 (as an upper limit to all well-mapped SNRs), distance from the Galactic plane, and brightness of the system (limited by the Eddington luminosity). But an identification of the Doppler-shifted emission-line *bullets* with the mapped outgoing radio knots yields the larger (and inherently noisy) distance of (4.85 ± 1)kpc. The discrepancy disappears when one drops the assumption of an orderly precessing straight-line motion of the jet material, in favour of channel-wall material dragged-along by the wiggly, corkscrew-shaped pair-plasma jets, see Chap. 11.

As already indicated, the controversy about SS 433 starts with the assumption of whether or not we deal with *soft beams*. If the moving emission

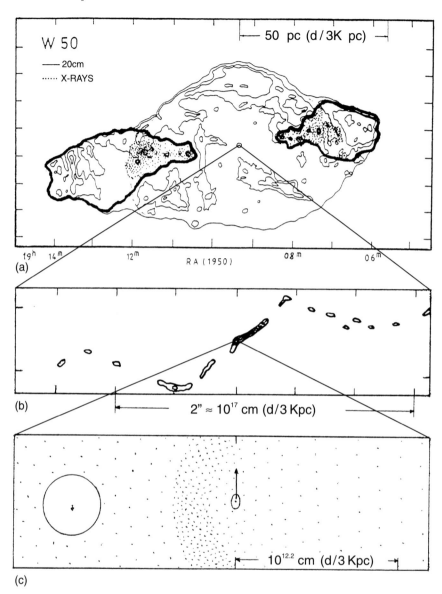

Fig. 13.2. (a) Schematic map of the old SNR W50 with its radio ansae (ears, lobes) and of its central jet source SS 433, probably a binary (ejecting) neutron star of age some 10^4yr. The central jets, seen enlarged in (**b**), are mapped at radio frequencies and traced (further in) as almost luminally moving emission lines, both at visible ($\lesssim 10^{15}$cm) and at X-ray ($\lesssim 10^{13}$cm) frequencies; their mild sidedness depends on a correct identification of their center. The jets blow the ansae, also traced as X-ray knots inside the confines of W50 proper (*stippled*). (**c**) The central binary system is unresolved, hence only predicted: A neutron star is required for blowing the (pair-plasma?) jets; a dense binary windzone is signalled by the strong and rapidly varying X-ray and optical emission lines, and by the strong and chaotic intensity fluctuations of the continuum. All emissions are multiply periodic, identified as orbital (13.0862±0.03 d) and precessional (164.0±4 d) motion plus several of their beat frequencies, including nodding

lines were emitted by hard beams, their power would have to be gigantic, $10^{42}C^{-1/2}$erg/s with $C :=$ clumpiness factor $\gtrsim 1$, instead of the otherwise bolometric luminosity of the system of only 10^{39}erg/s $(d/3\text{kpc})^2$. It would clash with the X-ray luminosity of W 50 where most of the beam power would have to be dumped, of only 10^{34}erg/s $(d/3\text{kpc})^2$ – which X-rays omit the ansae (!) – likewise with the ($\lesssim 10^2$-times stronger) IR luminosity from the dumping region. And the mapped jets are seen to be focused, rather than tracing the assumed precession cone.

To me, all the evidence points at SS 433 being a high-mass neutron-star binary whose 10^4yr-young neutron star precesses in magnetic interaction with its *inner (non-) accretion disk* and thereby liberates some 10^{36}erg/s of pair plasma (via reconnections, of Lorentz factor $10^{3\pm1}$) whose escape from the B-star's extended, orbiting windzone takes place in the form of two antipodal jets. The jets, in thrusting their escape channels, heat (to X-ray temperatures) and drag along confined clumps of wind material, looking like bullets of speed $0.26c$, which are radiatively ionized some $10^{6\pm1}$-times out to distances $\gtrsim 10^{14}$cm whilst they emit their observed recombination spectra. Sources similar to SS 433 are Cyg X-3 and the superluminal jet sources GRS 1915+105 and GRO J1655-40.

13.3 'Outflow' in Orion

The Orion Molecular Cloud and Nebula – M 42, NGC 1976, or *Orion A* – contains not only many bright young stars, among them the four (OB-type) trapezium stars, but also an outflow phenomenon with Hubble-flow kinematics traced by various masers, $\boldsymbol{v}(\boldsymbol{r}) \sim \boldsymbol{r}$, of range $\lesssim 0.5$ pc $(d/0.48\,\text{kpc})$, whose outflow center is the *Becklin–Neugebauer Kleinmann–Low* IR complex, some 0.12 pc NW of the trapezium (in projection). Due to $A_V \gg 12$ mag of visual extinction by the cloud, these fireworks are not recognised on optical maps and were only highlighted in the (northern) Summer of 1993 when > 20 ionized radial bullets trailed by conical hollow shafts had been mapped in [FeII] and molecular hydrogen, respectively. It was classified as a bipolar protostellar wind even though its kinematics and energetics are different, quite similar to the Crab, i.e. to a supernova whose outer two thirds do not show, see Kundt and Yar [1997: Ap&SpSc *254*, 1–12], and Plate 8.

The Hubble-flow kinematics – or SN kinematics – determine the *age* of this outflow through $v(r)/r \lesssim 1/10^{2.2}$yr as $10^{2.2\pm0.2}$yr; we thus deal with the youngest known SNR in the Galaxy, younger and more distant than Aschenbach's SNR RX J0852.0-4622 found at hard X-rays on the SE edge of the Vela remnant, at $d \approx 0.2$ kpc, age $\approx 10^{2.8}$yr [B. Aschenbach, 1998: Nature *396*, 141–144]. Its compact remnant is not yet identified: there are at least two suspicious candidates near the center of the Orion flow, within $5''$. The number of detected or inferred Galactic SNe during the past millennium has thus gone up to ≥ 9: Orion ($\lesssim$1840), Cas A (1680), Kepler (1604), Tycho

(1572), Kes 75 ($P/2\dot{P} = 0.72\,\mathrm{kyr}$), RX J0852-46 ($\approx$1320), 3C 58 (1181), Crab (1054), SN 1006.

13.4 CTB 80

CTB 80 tends to be classified as a SNR, like several other *exotic* SNRs which look like birds, like a rabbit, a tornado, a mouse, or otherwise exotic, whose spectra are similar to those of quasi-spherical SN shells but whose radiating electrons drain their energy from a fairly young pulsar rather than from their already faded supernova explosion, often at much lower power. I prefer to call them *pulsar nebulae*, or *synchrotron nebulae*, to avoid confusion with their younger, brighter, often more distant cousins [Kundt and Chang, 1992: Ap&SpSc *193*, 145-154]. Such a confusion may have entered the distance estimates of the SGRs – Sect. 10.2 – all of which may be nearer than 0.1 kpc.

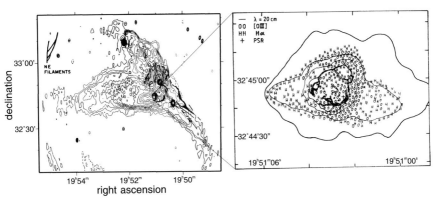

Fig. 13.3. Radio map (49 cm) of the Synchrotron Nebula CTB 80, blown by the $10^{5.0}$yr-old pulsar B1951+32, with a few Hα+[NII] filaments superposed. CTB 80 was formerly classified as a SNR, but the fading remains of the past supernova can hardly be seen any more, on IR and optical maps, encircling the shown radio source. The enlargement of the pulsar environment to the right has radio emission (20cm) as well as optical-line emission by [OIII] and Hα (denoted OO, HH, respectively) superposed. The pulsar is thought to drift towards the lower right, away from the center of the indicated spherical past windzone, at ($19^h51^m54^s$, $32°55'$)

CTB 80, the rabbit eating a carrot, does not look like a former windzone at all, see Fig. 13.3. Only after many years of devoted research, its $10^{5.0}$-yr-old *pulsar* B1951+32 was found in the eye of the rabbit, with an uncertain space velocity of $10^{1.8}$km/s ($d/1.6\,\mathrm{kpc}$) heading SW, towards the brightest edge of the pupil. When extrapolated backwards, it may have been born at the center of the vaguely indicated sphere, of radius $14' \simeq 7$ pc, which appears to be the progenitor's windzone. After 10^5yr, the still-glowing SN ejecta can barely be traced at IR frequencies, or even in the visible, as a large spherical shell around the system, of angular radius $1°$ (not shown in the figure). Most of CTB 80's luminosity, from radio through optical emission-line to X-ray, appears to be powered by the pulsar whose relativistic pair-plasma wind samples the regions of lowest density of its CSM at subsonic speeds, of order several 10^2km/s. Brightest is the expanding bipolar core of its windzone, the eye of the rabbit, of extent less than one lightyear.

13.5 Cyg X-1

Are there stellar-mass black holes in the Universe? When this suspicion first arose, some 30 years ago, the search was aimed at massive stellar X-ray binaries whose visible component oscillated significantly, forced by the orbiting gravity of its unseen companion. The prime candidate became Cyg X-1, an at least $20M_\odot$ O9.7 Iab star of orbital period 5.6 d, binary separation some $10^{12.5}$cm, at a distance of some 2 kpc, whose compact component had to have a mass in excess of 6 $M_\odot$. But a black hole is stationary, even when surrounded by an accretion disk with mass-transfer instabilities, whereas Cyg X-1 has revealed a lot of time-dependent structure that requires the additional presence of a neutron star in the system, with its hard surface inside a deep potential well (for the γ-rays), and with a corotating magnetosphere (for the variabilities, and relativistic wind generation). A small *radio jet* has been detected recently, see below. To me, the compact component of Cyg X-1 – as well as of all the other stellar-mass black-hole candidates – is a neutron star inside a *massive accretion disk*, typically of mass 5 $M_\odot$ [Kundt and Fischer, 1989: JAA *10*, 119-138; Kundt, 1999: MEMSAIT *70*, 1105–1112].

As already mentioned in Sect. 9.2, the list of stellar-mass *black-hole candidates* contains presently over 45 entries, five of them with *high-mass* companions ($\gtrsim 6$ $M_\odot$), the rest with *low-mass* ones ($\lesssim 2$ $M_\odot$). Their defining property is a mass in excess of 3 $M_\odot$ of their compact component. All the high-mass BHCs are persistent sources whereas most of the low-mass ones are transient, with recurrence times of decades. Every year, two or more X-ray novae join the list. Nevertheless – considering their much longer expected lifetimes (than those of the high-mass ones), by a factor of 10^3 – the low-mass BHCs are thought to belong to the rare variety among the BHCs, not to the representative one.

My suspicion of the BH interpretation comes from (i) a number of spectral and lightcurve properties which require a *hard surface*, an *oblique magnetic dipole*, and two dense, *interacting windzones*; (ii) the *indistinguishability*, as a class, of the BHCs from the neutron-star binaries in all properties other than their inferred mass; and (iii) the missing *intermediate-mass systems* which should naturally evolve into neutron-star binaries with massive disks.

Concerning properties (i), and starting at *γ-ray energies* near MeV where Cyg X-1 and other BHCs have a significant – or even dominant – radiation output during their X-ray soft state: A well-fed several-$M_\odot$ black hole should not radiate much above its equipartition temperature, of order keV; the hard surface of an accreting neutron star is required for 10^3-times higher temperatures. Proceeding down to *X-rays*, why do the lightcurves of the transient sources saturate at $L_{Edd}(1.4\ M_\odot)$, even during outburst, rather than at the hole's (much larger) Eddington luminosity? And why do their lightcurves drop exponentially after outburst, within $\lesssim 2$ yr, to an X-ray *quiescence* level of (only!) $10^{32\pm2}$erg/s? This is impossible for a standard, low-mass accretion disk; a rigidly rotating, non-dissipative disk seems to be indicated. Next, at *visible frequencies*, we encounter noisy, non-periodic lightcurves whose stacked averages lack reflection symmetry w.r.t. their two orbital minima – a property shared by systems with a symmetry plane (through the centers of the two components) – and strong linear polarization ($10^{-2.5}$) and circular polarization ($10^{-3.3}$) in the case of Cyg X-1, and with fluctuating, broad emission lines; two interacting, magnetized windzones appear to be present. Finally, at radio frequencies, the repeated *radio outbursts* are reminiscent of relativistic-pair formation in strongly reconnecting magnetic fields.

(ii) Among the long list of remarkable properties in which the BHCs are indistinguishable, as a class, from neutron-star binaries are (j) the presence of a *third (precessional) period* of several months, 294 d in the case of Cyg X-1; (jj) a hard-soft state *spectral bimodality*, pivoting around 6 keV, and extending up to MeV; (jjj) their *flickering*, expressed by their X-ray *power spectra* which range from mHz to $\gtrsim$ kHz and show various *quasi-periods*, in particular of several 10^2Hz, up to 1.2 kHz, reminiscent of innermost Kepler periods, of a spin period, and/or of beat frequencies thereof; (jv) their *jet-formation* capability; (v) their occasional *super-Eddington* X-ray luminosities (requiring feeding by a massive disk?); (vj) quasi-periodic *X-ray dipping* (caused by the neutron star's wind?); (vjj) *type II X-ray bursting* (via accretional instabilities); (vjjj) *polarized* optical lightcurves (suggesting ordered magnetic fields), (jx) *Li* in absorption (unlike in cataclysmic variables; formed at the neutron-star surface?); (x) transient orbital-period variations (by $(1\pm0.5)\%$) during *superhump* state, familiar from the SU UMa class of *dwarf novae*, and to be explained by illuminated clumps orbiting beyond the outer edge of the disk.

(iii) As indicated above, compact binary systems with *heavy accretion disks* are expected to form from close progenitor systems in the *intermediate-mass* range, $(6+6)M_\odot$, say, in which the more massive component evolves faster than its companion, transfers part of its mass through its wind, goes SN and leaves a neutron star behind which subsequently accretes mass from the originally less-massive component, during the second stage of (reverse) mass transfer. For typical transfer rates of $\lesssim 10^{-5}M_\odot$/yr, most of this matter must stay in orbit around the neutron star because of its Eddington constraint (6.12), and will redistribute its specific angular momentum towards that of a rigidly rotating McLaurin spheroid. During this phase, the growing disk, of mass $\lesssim 5\ M_\odot$, should shine as a bright, *supersoft X-ray* source, see Sect. 6.2.

During later stages, this massive, degenerate disk can give rise to both super-Eddington luminosities, and to BH candidacy.

Returning to Cyg X-1 as one of the brightest and best-known BH candidates – which has almost all the remarkable properties (j) through (x) above – is there a direct proof of a neutron star inside its compact component? I do not think so. Clearly, when the neutron star wants to blow *jets*, most of its pair-plasma gets destroyed in situ by inverse-Compton losses on the photon flood from its near companion, dominantly so on the side facing it. But a faint, 1-sided, 15marcsec long radio jet has been recently detected with the VLBI. Short, 1-sided radio jets have been likewise detected in the high-mass X-ray binaries LSI+61°303 and LS 5039, hence are likely to be a common phenomenon, not only of the low-mass systems. Screened neutron stars are likely to hide in all the non-white-dwarf X-ray binaries, but are difficult to identify.

13.6 Eta Carinae

One of the brightest stars in the Milky Way and Local Group (of galaxies) is η Carinae. During the middle of the 19th century, it underwent an outburst whose integrated light equalled that of a supernova. This and another outburst near the end of that century launched the rapidly expanding *homunculus nebula* in which η Car is embedded, whose morphology is unlike any other emission nebula in the sky, see Plate 8. At an estimated distance of (2.3 ± 0.2)kpc, its bolometric luminosity equals $10^{40.3}$erg/s $= 10^{6.7}L_\odot$ of which only a small percentage escapes in the visible. This small percentage has started to rise near the middle of the last century, from less than 1% to a present 10%, and may well continue to do so; does the fog clear, via expansion? The kinematics of the homunculus can be described by two approximate Hubble flows launched in $\{1844 \pm 7, 1885.8 \pm 6.5\}$, at speeds of $\lesssim \{10^{3.5}, 10^{3.1}\}$km/s.

η Car has recently revealed two periodicities: 5.53 yr and 85 d, from radio through X-ray frequencies. The longer of them has been interpreted as the highly eccentric motion, $\epsilon \approx 2/3$, of a (hotter, primary) B2 Ia star around a more massive (secondary) B8 Ia star, with a semi-major axis of $\gtrsim 9$ AU, whereby the total mass in the system remains ill-determined, between $40\,M_\odot$ and $10^{2.5}M_\odot$. High masses would be required by a strict Eddington limit (6.13), in particular during the 19th century, whereas much lower masses are suggested by stellar statistics. In the latter case, the super-Eddington outburst can be blamed on an additional neutron star in the system [Kundt and Hillemanns, 2002, MEMSAIT, submitted].

A third component in the η Car system is indicated by periodic *X-ray peaks* of roughly constant relative intensity, at intervals of 85 d (and in between), even though this period is not clean, and involves higher harmonics. Repeated radio maps show *luminal expansions* at super-thermal brightness temperatures, reminiscent of pair-plasma injections. These constraints can be met by a neutron star in close orbit around the secondary component, with a semi-major axis of some 2 AU. Neutron-star binaries of this separation tend to be highly eccentric; and details of the X-ray lightcurve, in particular its sharp orbital minimum, suggest a highly eccentric orbit for the neutron star. Neutron-star binaries are known to have occasional *super-Eddington* X-ray outputs, with luminosities $\lesssim 10^{41}$erg/s, probably when accreting from heavy disks; see Sects. 9.1 and 13.5. Rather than making one of the non-degenerate stars responsible for the outbursts in the 19th century, on time scales much shorter than the thermal time scales E_{th}/L of their envelopes, one may blame the burden of the transiently huge luminosity and violent *mass ejections* on the unseen neutron star in the system, during epochs of disk reconfiguration. At the same time, the strange morphology of the homunculus would find a rare explanation: as driven mainly by relativistic pistons, radiation pressure and pair plasma.

13.7 Sgr A

An astrophysical object of key concern is the center of the Milky Way, Sagittarius A*, at a distance of $\lesssim 8$ kpc, hidden behind $(31\,^{+7}_{-4})$ magnitudes of visible extinction, a dimming by a factor of $10^{-12.4\,^{-2.8}_{+1.6}}$. We know from Sects. 6.2 and 9.2 that galactic disks have much higher densities, pressures, and rotation velocities in their innermost 0.1 kpc than elsewhere; so what do we find near the center of our own Galaxy? If its disk was strictly plane, we would not be able to see anything. But thanks to its warps (by $\lesssim 25°$), starting with Gould's belt at $\lesssim 0.5$ kpc and dipping through the plane at least once, we have a chance, at wavelengths $\gtrsim$ μm and $\lesssim$ nm ($\gtrsim$ keV), to see the center so-to-speak from above, from the side of the inner Gould belt where Galactic rotation appears clockwise.

   The Galactic rotation curve tells us that at well-resolved distances, $v(r) \approx \sqrt{GM(r)/r} \sim r^{0.1}$ holds, whence $M(r) \sim r^{1.2}$, $\rho(r) \sim r^{-1.8}$. Average *gas densities* and pressures are thus expected some 10^4-times higher inside the Galactic-center *chimney*, a cylinder of radius (50 ± 20)pc (depending on direction) which intersects the Galactic disk at right angles and whose edges are marked by *fibrous*, or *filamentary threads* of non-thermal radio emission. In my interpretation, this chimney is the escape channel for pair plasma of Lorentz factor $\gamma \approx 10^4$, generated by the central engine*Sgr A**, whose escape into the halo can be traced through more than 8 kpc, in the form of two mapped 21-cm knotty jets which merge, beyond some 3 kpc, into the ribbon of falling *high-velocity clouds* in the halo [Kundt, 1990b, 1996]. These old jet walls would tell us that our Galaxy has shown Seyfert activity in the past, some 10^7yr ago, like all the massive galaxies with gaseous disks; see Plate 9 and Figs. 13.4 and 13.5. The thermal walls of the chimney, of thickness

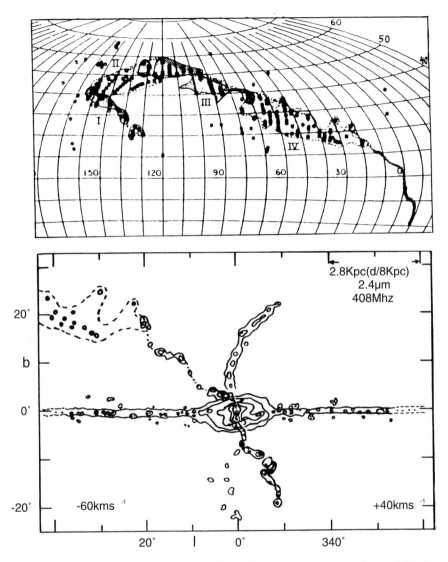

Fig. 13.4. Homing in on the Center of our Galaxy, at various radio and IR frequencies and in Galactic (l, b) orientation, starting on the Galactic scale (of 8 kpc) with the falling high-velocity clouds (in the northern-Galactic hemisphere), and following them all the way in to the innermost fraction of a lightyear, dominated by the radio point-source Sgr A*, in Fig. 13.5. Remarkable are the approaching (-60 km/s) and receding $(+40 \text{ km/s})$ jet, apparently forming on the pc scale from within Sgr A West and Sgr A East, and the stack of coaxial cylinders surrounding them on the $\lesssim$ kpc scale, perhaps recording earlier active epochs of our Galactic Nucleus. See also Plate 9, and [Kundt, 1990b, 1996]

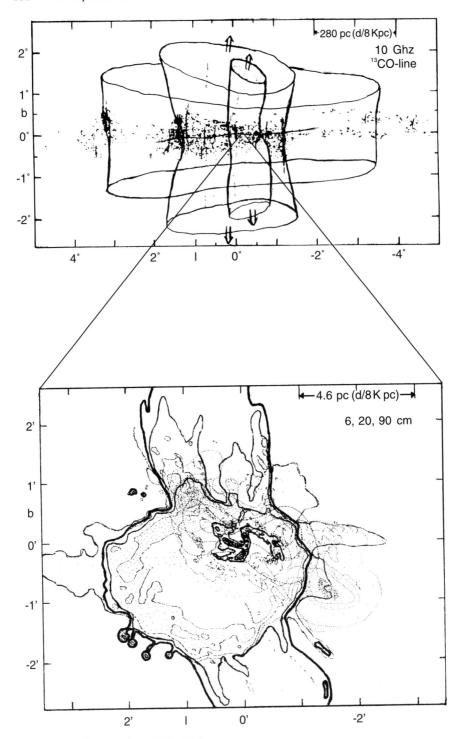

Fig. 13.5. See caption of Fig 13.4

some 1%, consist partially of matter at *forbidden* (non-corotating) *velocities* which may have been transiently dredged up by the escaping jet substance, and/or condensed on them from outside, and is falling back into the disk, thus realising a huge (dynamic) scaleheight.

On further approach of the Galactic rotation center, we meet the non-thermal radio source *Sgr A East*, an elliptical shell of semi-major axis 3.7 pc one quadrant of which is occupied – at least in projection – by the thermal *vortex Sgr A West*, also called *mini-spiral*, or *triskelian*, which in turn harbours the rotation center *Sgr A**, a radio quasi-point-source of luminosity $10^{2.5} L_\odot$, extent $\lesssim 3$ AU, perhaps elongated by a factor of $\gtrsim 3$. Its position has the unrivalled accuracy $(\alpha, \delta)(2000) = (17^h 45^m 40.0383^s, -29°00'28.069'')$. Its mass has been determined from the orbits of $\gtrsim 10^2$ nearby stars as $M = 10^{6.46 \pm 0.05} M_\odot$. It emits $\{10^{2.5}, < 10^{1.5}, \leq 10^{4.7}, 1, 10^{3.1}\} L_\odot$ at $\{\lesssim 10^{3.5} \text{GHz}, 10^{14.1} \text{Hz}, \approx 10^{15} \text{Hz}, \text{keV}, \gtrsim \text{GeV}\}$, whereby the entry at $\gtrsim$GeV energies has a positional uncertainty and extent of $0.2°$, hence may come from anywhere inside the chimney; but it is unique in the Galaxy, hence certainly related to *Sgr A** [Mayer-Hasselwander et al., 1998: Astron. Astrophys. *335*, 161–172]. There is a recent claim that the radio lightcurve of *Sgr A** shows a 106-d periodicity.

To understand the functioning of this galactic nucleus, it is fundamental to know the relative locations of its components, not just in projection. In [1990b] I explained why I consider all data consistent with the *containment* relation *Sgr A** $\subset$ *Sgr A West* $\subset$ *Sgr A East* $\subset$ *chimneys*, which allows *Sgr A** to be the source of relativistic pair plasma which has blown both *Sgr A East* and all the chimneys. This re-interpretation of the wealth of data involves the claim that *Sgr A East* is $10^{1.6}$-times more energetic, and $\lesssim 10^{-1}$-times smaller than a *supernova remnant*, and is associated with peripheral compact HII regionswhich would mismatch in age; it may serve as a *storage bubble* for the Blandford and Rees deLaval nozzle of the Galactic twin-jet which can be seen to take off from *Sgr A East* to both sides of the disk. Second, it involves the re-interpretation of continuum absorption as a lack of emission, below the density-dependent lower (Razin) cutoff at $\nu_R = (\gamma \nu_p^3 / \nu_B)^{1/2} = 10^{8.7} \text{Hz}(\gamma_4 n_4^{3/2} / B_{-2})^{1/2}$ which grows monotonically from *Sgr A East* through *Sgr A West* to *Sgr A**. All spectral and morphological details of the *Sgr A* complex support this interpretation. A recent X-ray map by CHANDRA has confirmed it.

Once we try to identify *Sgr A** with the cause of all the unusual Galactic-center phenomena, we must explain its functioning. Its feeding rate via *spiral-in* through the disk has been variously estimated at $10^{-2.5 \pm 1} M_\odot/\text{yr}$, equal in magnitude to the blown-out *wind* which manifests itself through the cometary tails of ≥ 5 nearby stars. The 20 $M_\odot$ of *Sgr A West* are next in the accretion queue. An engine of mass $10^{6.5} M_\odot$, fed at a steady rate of $10^{-2} M_\odot/\text{yr}$, can radiate $10^{42} \epsilon_{-3}$ erg/s $= 10^{8.4} L_\odot$ at an assumed efficiency of $\epsilon = 10^{-3}$, without any restriction by the Eddington limit. But as mentioned above, the present bolometric luminosity of $Sgr\ A^*$ is only $\leq 10^{4.7}L_\odot$, embarrassingly low for a black hole whose feeding would have to be $\leq 10^{-6.2}$ sub-Eddington. A burning disk, on the other hand, can shut off transiently to very low luminosities, by rotating almost rigidly in ring-shaped domains.

Can we estimate the output of $Sgr\ A^*$, better than was done empirically above? Its power-law radio spectrum of spectral index $\alpha = 0.3$, with upper cutoff near $10^{12.5}$Hz, suggests *monoenergetic synchrotron* radiation with $\gamma^2 B_\perp = 10^{5.9}$G, see (3.10). From this and seven further constraints, I once predicted $\gamma \gtrsim 10^4$ [1990b]; which was subsequently corroborated by new measurements, both on the *arc* region of the chimney, and on the young synchrotron jet. In the meantime, some of the earlier input data have changed, such as the point-source fraction of the pair-annihilation power, and younger workers have preferred higher estimates of the magnetic-field strength in the emission region, thus lowering the estimate of γ. But they ignore the hard γ-rays which may well be the inverse-Compton losses of the escaping pairs on the central photon bath, of peak energy 3 GeV, which require $\gamma \lesssim 10^{4.5}$. As these losses are estimated to be a small percentage, the emitted *power in relativistic pairs* should be distinctly larger, above 10^{38}erg/s. Independently, if all the pairs annihilating in the central bulge (mapped by their 511-keV line), $\dot{N}_{e\pm} = 10^{43}$s^{-1}, were once released by $Sgr\ A^*$ with $\gamma \gtrsim 10^4$, the inferred pair power rises to 10^{41}erg/s $=10^{7.4}L_\odot$; which may, however, apply to a past era when our nucleus was more active than today. As a rule, the bolometric power of a radio-loud AGN tends to be 10^2-times its relativistic-wind power; why is it not seen?

Another property which highlights the non-thermal aspect of $Sgr\ A^*$ is the high percentage of its *circular polarization*, rising above 1% beyond 10 GHz, and of its *linear polarization*, rising even above 10% beyond 150 GHz. Do they reflect a high order of the magnetic fields in the emission region, whose reconnections give birth to the relativistic pairs? Is the rise of the polarization at high frequencies a signature of the transition from optically thick to thin?

14. Astrobiology

Life on Earth, as realised by plants and animals including man, tends to be often better organised than civilization. The *biological engines* are so stable and effective that they have managed to survive throughout several Gyr, despite all natural hardships, despite being eaten mutually and/or hierarchically, and despite all natural catastrophes. Even more: until the past century, the *diversity* of species has grown roughly *exponentially* with time – interrupted by some ten mass extinctions – thereby conquering all thinkable *niches*: sweet and salt water including oceanic depths (of $\lesssim$ 10 km), land and air, extreme (high and low) temperatures and pressures, corresponding to depths and heights of $\lesssim$ 10 km, salt- and sulfur-rich media, darkness and dryness, and even as toxic environments as the interiors of stomachs, or certain locked-off caverns. When they say that life has had more than a Gyr to adjust, they forget that earlier generations had to survive as well, in order to hand over the torch of life: Life must have been well adjusted right from the beginning, down to its most primitive forms. And Darwinian *survival of the fittest* is a mechanism that may well work at the level of one-celled creatures – which duplicate within 20 min, and exist in huge numbers – but not of lions, elephants, eagles, and dolphins for whose progenitors there are simply not enough generations for random selection; Gold [1999] calls it '*Darwin's dilemma*'. Biology is a miracle, barely explainable by the *anthropic principle* – according to which the whole Universe, including the laws of physics, has been devised to allow for at least one habitated planet (or moon). This chapter attempts to highlight the building plan, and high degree of technical perfection of all living beings with their dynamical skills, differing chemical factories, large numbers of senses and sensitive feedback circuits, high efficiencies, sophisticated symbiosis, and intriguing evolution. Is life a *cosmic imperative* [Duve, 1994]?

Those who will find this brief chapter stimulating may wish to deepen their knowledge. To them, I recommend the books by Sergeev [1978], McMahon and Bonner [1983], Forsyth [1986], Dröscher [1991], and Varju [1998] on phenomenology, by Hoyle [1975, Chap. 12] as an introduction to its chemistry, by Dyson [1985] on the probable double origin of life, by Gold [1999] on a possible underground origin, by Layzer [1990] on cosmology and biophysics, and by Barrow and Tipler [1986] on the anthropic principle. Life takes place on Earth, and even abiogenic Earth holds its uncounted secrets; Walker [1977] collects over 600 of them, including answers and references.

14.1 Examples of Life

Man is used to coexisting with animals, partially to feed on them, and partially to even be helped by them: *horses, dogs, oxen, yaks, camels, elephants, doves, falcons, dolphins*, and several other animals have cooperated with man in past centuries, performing work, keeping guard, tracing, hunting, or conveying news. Animals could help man because they were stronger, faster, more sensitive, or better equipped to the climate even though they were rarely more skillful, and even though the only mode of communication was gestures and sounds. We are also used to coexisting with plants of all sorts, mostly for feeding, but also for heating, constructions, shelter, drugs, and comfort. Plants serve us even though we often treat them at will. Is all this consistent with the laws of physics? At least we have no hint that *life* could violate any law of physics, except possibly by its remarkable stability.

We often take for granted that certain animals have abilities clearly superior to us. For instance, *cats* race through darkness at speeds unimitable by man. Their eyes are proverbial optical mirrors, achieved by a reflecting *tapetum lucidum* between retina and choroid (= layer of veins); for the purpose of radiative cooling? *Owls* even fly through the dark. They have sensitive ears; but how do they manage to control their nightly flights without active radar? Are the eyes of nocturnal predators far-IR sensitive?

Camels can work in the desert for over a month without drinking. They can store fifteen pails of water in their body, in their (swelling) red blood corpuscles, and they do not use it for cooling (sweating); they allow their body temperature to drift, between $35°C$ and $40.5°C$. For this they have various enzymes which change guards on transition to adjacent temperature intervals. And they recycle their urine in a rumen (= first stomach). For better cooling, they store their fat not under the skin but in their humpbacks. And they produce water by burning fat. A recently discovered Tibetian species of camels can even drink salt water. In contrast, *water spiders* build caissons under the surface, attached to water plants, and fill them with air bubbles carried between their hind legs; they can thus breathe under the surface whereby diffusion guarantees a stable oxygen percentage of 16% during moderate consumption.

Another extraordinarily equipped animal is the *sperm whale* whose meals (squids) entice at the bottom of the oceans, at depths of 3 km and deeper. Sperm whales get their shuttling for free, by sinking like stones, overtaking diving submarines, and resurfing via buoyancy, like air bubbles. For this to be feasible, they lower their body temperature gently before diving, via countercurrent heat-exchange blood flow through their fins and via strong breathing for at least a quarter of an hour, store half of the oxygen in the myoglobin of their muscles and tissues, then short-circuit their blood circulation to stop cooling (by storing most of the blood in their 'Wundernetze'), exhale the nitrogen, collapse their lungs, and fall head first, pulled by the ton of wax in their heads which freezes at 36°C and thereby shrinks considerably, and by the cooled oil along their backbones whose thermal expansion coefficient exceeds that of water by $\gtrsim$ 10%, thus pushing the whale's specific weight distinctly above that of the ambient water. Prepared like this, the animal's body temperature starts rising again, in proportion to its burned oxygen, because its massive body is well insulated by the fat (blubber) under its skin. And with the reliability of an hourglass, the wax eventually melts and pulls its head back to the surface, after some 1.5 h, in time for the next breath.

Among the specially equipped animals are the *giraffe* whose height (of $\lesssim$ 6 m) approaches that of a blood-pressure scaleheight and would cause problems of changing pressure in its brain and feet, had not its veins been provided with safety valves and contractile segments, assisting their heart. Next think of *salt-water fish* which avail themselves of distilling glands to reduce the salt concentration of their blood – as do sea birds and certain reptiles – in order to keep the osmotic pressure in their bodies near the physiological 6 bar, distinctly below that of the sea. Think of *ruminants*, like camels, cows, giraffes, goats, and sheep that have an extra stomach filled with bacteria which allows them to digest cellulose, which other animals (and man) cannot recycle. Or think of *wallabies* (and *kangaroos*) which can transiently store kinetic energy elastically in their long feet – in the long tendons and/or bones – in a certain load range such that the running speed of a mother is not reduced when carrying a baby, at non-enhanced effort (measured by oxygen consumption).

Animals can solve problems of orientation better than unequipped man: Imagine birds or fish which can migrate around the globe and find their way back to the tree or pool of their breeding place, timed to the day of the year. Extreme cases are *doves* which may use all the senses we can think of to find their way home: eyesight (polarized), sun and/or stars, odours, and magnetic fields (by sensing the Lorentz force with a special organ in their beak?), perhaps even coarse IR (via the comb organ in their eyes, [Dröscher, 1991]). Clearly, magnetic-field directions should be quite unimportant to their needs, or status of life – contrary to odours – but they may help keeping a chosen flight direction. We still do not know for sure. During their long passages, in addition to burning their fat, migrating birds can transiently recycle parts of their unused organs (liver and kidneys). Even more impressive is the habit – and ability! – of certain kinds of *salmon* to smell their way home from the Sargasso Sea, at an age of seven, to the northern European pool of their birth, upstream against rapids and water falls, never missing the correct river branching, irrespective even of hungry bears along the banks.

A requirement for survival in cold climates is *hibernation*. Methods to survive through low temperatures, all the way down to absolute zero, are

(i) *desiccation* (of spores, seeds, or even higher organisms), (ii) lowering of the freezing point or crystallisation by solutes {glycoproteins, sugar and alcohol} below {−1.9, −9}°C, (iii) supercooling (by slow circulation?), and (iv) controlled periodic reheating. Among the vertebrates, they are variously applied by certain squirrels, frogs, turtles, bears, fishes, and birds. Another form of conservation is the ability of several insects to *anaesthetise* their prey for weeks or months by injecting narcotics, in order to provide live food for their offspring.

Astonishing achievements by animals are said to be performed via *instinct* – like the knots made by certain birds with their beaks during nest construction, the incubators of the mound birds formed from rotting plant debris which are temperature-controlled to 33°C, the nets knitted by spiders, the widespread building and finding again of separate food reservoirs, or the coordinated constructions and entertainments of temperature-controlled shelters by colonies of bees or ants. Are such achievements qualitatively different from those by man, or had we not better speak of *biological intelligence*, in characterization of non-random processes steered by neural nets? Is the functioning of human babies qualitatively different from that of animals? Such *mental* steering appears to be both universal and essential for the functioning of life.

Plants and animals show various forms of cooperation known as *symbiosis*, in all possible combinations: *Mycorrhyzae* (fungi) serve as underground merchants, trading water and minerals with different species of trees for the products of photosynthesis. *Fertilization* of blossoms by bees, flies, spiders, butterflies, birds, bats is more or less evenly spread throughout the year, apparently for optimization, steered by odours, colours, baits, or acoustic reflectors. *Eucalyptus trees* are the lifetime asylum for *koalas* – whilst poisonous for most other creatures – and *mangrovae* feed *nose monkeys*. *Acacias* feed certain *ants* which protect them against giraffes. Attini *ants* cultivate certain *fungi* and an antibiotic-producing *bacterium* which protects the fungal gardens (since some 50 Myr, [Nature, 2000: *398*, 747]). The African *honeyguide* (bird) feeds on bees' wax for whose access it accepts the help of some stronger robber, possibly man. The *ice bear* hunts jointly with a *polar fox* and a *bird*, for mutual benefit.

Even more sophisticated – and deterring – are the survival strategies of parasites like the *lancet fluke* (= Hirnwurm, kleiner Leberegel) which lives in the liver of farm animals (cows, sheep, but also deer, marmot, man), later leaves its temporary host's stomach in the form of eggs which are eaten by *snails*, evolve into larvae, then sporozystes and via daughters into cercariae which traverse the snail's body, are coughed out and subsequently eaten by *ants* (in confusion with eggs), in whose bodies they develop into metacercariae which reset their nervous system such that when it falls cooler, an infected ant climbs up a blade where it waits, in a torpid state, to be eaten by a farm animal in order to supply its inherent metacercariae to this animal's inner organs where they can close the cycle, in becoming new lancet flukes.

Comparably deterring is the dinoflagellat *pfiesteria*, a monocellular alga which has made its first appearance in American estuaries in 1988. Once every few years, it has killed and digested millions of small fish, typically in less than a day. This killer microbe can take at least 24 different shapes, of various forms of amoeba, cyst, zoospore, gamete, and planozygote of which only the toxic version of the zoospore is the poisonous member. Its toxin attacks the nervous system, and is still effective in the air above a haunted water basin [Burkholder, J.-A.M., 1999: Scientific American *281*, August, 28–35].

This minute selection of examples of biological optimization from within the $\lesssim 10^{7.7}$ species living at the surface of present-day Earth may suffice to introduce the players and to explain what is meant by the wonder of biological efficiency: how can the biosphere keep its entropy thus low?

14.2 Water

All *transports* in the biosphere use water as the convecting liquid, both in the form of *sap* in plants, and in the form of *blood* in all higher animals. Should this be considered an accident, or an essential for life in the Universe? This section will remind us of twelve properties of *water* in which it is superior for life to all other liquids. It is thus hard to imagine that life anywhere in the Universe could do without water.

1. When compared with other liquids like CH_4, NH_3, HF, H_2S, HCl, oils, benzenes, or the noble gases, H_2O is by far the most *abundant* molecule on Earth and also in the Universe, allowing for oceans, frequent rain, rivers, lakes, and glaciers.

2. The liquid *temperature* range of water at terrestrial surface pressures, $273 <$T/K< 373, is distinctly at higher temperatures than that of all the other substances, by typically 10^2K, and corresponds to particle kinetic energies of 0.03K which allow for the relevant chemical reactions to take place in reasonably short times. Note that even for water-based life, Homo Sapiens took almost the whole 10^{10}yr of the Sun's main-sequence burning stage to be created.

3. Water does not *burn*, unlike e.g. CH_4 or oils, so that woods are not easily destroyed by fire. This threat may grow more severe in the future $10^{2.3}$Myr, because of the continuing hydrogen loss from the exosphere and consequent oxygen enrichment of the atmosphere.

4. The *density anomaly* (of 9%) of water near its freezing point (for pressures $\lesssim 2.2$ kbar) guarantees that lakes hardly freeze all the way to their bottom, and that glaciers flow (on a film of water). It also allows ice to split rocks, and seals to build their caverns in floating ice, above sea level. This rare property

is only shared with Sb and Bi; it is due to the bonding angle of $104.5°$ at which the two H atoms are seen by the oxygen atom, which is close to, but different from the angle $2\arcsin\sqrt{2/3} = 109.5°$ at which the midpoint of a tetrahedron sees two of its vertices.

5. With a pH value of 7, water is chemically *neutral*, unlike several of the comparison fluids.

6. Its large *dielectric* constant ($\epsilon = 81$) allows water to dissolve a large number of substances; such a high ϵ is only exceeded by few liquids, among them H_2SO_4, HF, and HCN.

7. Water has the highest *heat of vaporisation*, making it an ideal coolant.

8. Its high *binding energy* to oxygen makes hydrogen in air the second best fuel, rivalled only by HF. Water is thus naturally produced during energetic metabolism.

9. The high *surface tension* of water – exceeded only by Se – helps in particular the leaves of plants to secure their water against losses by gravity and evaporation.

10. *Cohesive* forces of water are so high that under ideal conditions, thin liquid threads do not tear under gravity for vertical lengths up to 3 km (measured centrifugally, with a bent, rotating tube).

11. Water is sufficiently *viscous* to slow down the thermal random motions of large molecules, enough for high reaction rates.

12. Water is *transparent* in the visible but *opaque* at far IR frequencies, allowing for daytime greenhouse heating of (surface layers of) oceans, lakes, and rivers.

14.3 Essentials for Life

Having argued that water is the ideal *transport* medium for life, let us next look at the further building materials of organisms which guarantee *rigidity* and *mobility*, a *reproducible* structure, and fast *supply* during construction and maintenance. Lehninger has classified 6 chemical elements as of *class I* (H, C, N, O, P, S), another 5 as of *class II* (Na, K, Mg, Ca, Cl), and additional 16 trace elements as of *class III* (Mn, Fe, Co, Cu, Zn, B, Al, V, Mo, I, Si, Sn, Ni, Cr, F, Se), all of which are essential in the nutrition of at least one species. Man may even require more than ten additional elements for rare but essential purposes (in enzymes). Apparently, life makes use of a significant fraction of all the chemical elements, all of which happen to be available at the surface of Earth thanks to the stirring action of deep-rooted volcanism (including plate tectonics) whose feeding plumes (tubes) drain on its fluid core. And a close look at all the various molecules used in plants and animals leaves me with the impression that *organic chemistry* is essentially the science of the *biological building blocks*, which function like tinker toys, or rather like LEGO blocks.

A leading role in organic chemistry is played by the element *carbon*, for which no substitute is readily in sight: Its central position in the periodic table of elements allows it to bind strongly to itself, forming chains, rings, and tubes, as well as to various other atoms. Its alloys outnumber those of all other elements. Without carbon, life would hardly find enough building blocks to choose from. Its nearest rival is silicon whose corresponding hydrogen alloys are so weakly bound that their chemistry would require temperatures too low for reasonable reaction rates. Living organisms consist to 71% by weight of water, and to 20% of carbon, another 7% being almost equally shared by Ca, P, and (excess) H. CO_2, the burning product of carbon, is a gas of which man exhales a kg per day. It has an anomalously high solvability in water, in violation of Dalton's law, and is an efficient *buffer* of the acidity (pH value) of a solution.

The basic *building blocks* of living organisms are autonomous *cells* which commit suicide (apoptosis) when malfunctioning. They are separated from their surroundings by at least one insulating, hydrophilic, phospholipid *membrane*, $\gtrsim 10^{-2.3}\mu$m thick, consisting of two adjacent layers of (hydrophobic) fatty acids topped on both sides by phosphate heads. These membranes contain many protein sluices for exchanges with the outside world, specific for each type of cell. For pulling big molecules (such as glucose) through a watertight sluice, an electric voltage of $10^{-1.15\pm0.15}$V is provided by permanently active *Na-K pumps* which suck Na^+ ions to the outside in order to charge it positive. Na^+ions can thus serve as electric engines pulling cargoes electrostatically to the interior through narrow channels. The pumps are thought to act as heat pumps, arresting the Na^+ ions on the outside once they have managed to thermally bounce across the opposing voltage [Kundt, 1998b]. For this to be possible without violating the second law, the gate keeper has to be powered each time by an ATP molecule (= adenosin triphosphate) – the biological energy unit (= 0.32 eV) – which is hydrolysed in this process to ADP (plus H_3PO_4). Of course, steady-state operation requires a power station inside the cell – the *mitochondrion* – which recycles ADP to ATP. No work is performed in compartments lacking a mitochondrion, or some similar organ, like a chloroplast. Note that a high thermodynamic efficiency (of order 0.66) is required for all the cellular engines in order to avoid overheating during action. Multicellular life has probably used them right from the beginning.

Ion pumps similar to the $Na-K$ pumps, common to all plant and animal cells, have diffusive charging times of order ms, which sets a lower limit to biological reaction times. They are also used by some fish for *orientation* in turbid waters: such fish emit electric pulses of several V, obtained by stacking several batteries in series, and map with their lateral-line organ (so to speak) the mirror charges in their surrounding conductors, being sensitive to field strengths of $\gtrsim 5$ nV/cm. Much higher voltages, of up to 0.8 kV, are generated by the *torpedo ray*, the South-American *electric eel*, and the African *catfish* in order to paralyse their prey, by stacking some 10^4 batteries in series; their discharge power peaks at 10 kW, for a small fraction of a second.

Returning to *cells* as the building blocks of higher organisms, it should be mentioned that their static solidity is achieved by means of *rods* and *ropes* crossing their interiors (as with tents), the rods consisting of *tubulin* tubes, the ropes of *actin* filaments. Ordered transport inside cells takes place along these 1-d structures, via molecular engines, viz. via dynein and myosin motors which slide along them, reminiscent of conveyor belts at airports. Even inside individual cells, traffic is not left to diffusion! Moreover, the higher rigidity requirements of plants (compared with animals) are met by additional (primary and secondary) *cell walls*, formed primarily from cellulose fibers, of thickness $\lesssim 0.3$ μm. These walls have to take pressures of $\lesssim$ bar, exerted by stepwise jumps in the osmotic pressure, a phenomenon known as turgor. During cell growth (through factors of $\lesssim 10^6$ in volume), these strong walls are distended by transiently loosening the cohesive network of polysaccharides (cellulose microfibrils) via the action of *expansin* proteins [Nature, 2000: *407*, 321]. The number of cells of an organism grows with its size: {fungi, plants, mammals} consist of $\lesssim \{10^{12}, 10^{15}, 10^{16.4}\}$ cells, of some $\{4, 20, 10^2\}$ different types, respectively.

There is no present-day organism without a building plan: in 1953, Watson and Crick discovered the structure of this carrier of information: a double helix called DNA ($=$ *deoxyribose nucleic acid*) whose rails (strands) are formed alternatingly from sugar and phosphate, and whose ties (rungs), stretching between successive sugar links, are formed from the two base pairs AT and GC of nucleic acids called adenine, thymine, guanine, and cytosine, respectively, which are linked by (weak) {double, triple} bindings of equal total length (20 Å), comparable to a zipper [Hoyle, 1975]. Whilst the strands consist of carbohydrates and phosphorus, the rungs require nitrogen. DNAs of {viruses, bacteria, fungi, plants, insects, amphibians, mammals (including man)} have lengths of $\{\lesssim 10^4, 10^{6.5\pm0.5}, 10^{7.5\pm0.5}, 10^{9.5\pm1.5}, 10^{9\pm1}, 10^{10\pm1}, 10^{9.6\pm0.4}\}$ rungs. Rungs have separations of 3.4 Å, resulting in a total length of a human DNA of $10^{9.8+0.5}$Å $= 2$ m. This giant DNA molecule, with the topology of a ladder or railroad track endowed with an intrinsic direction, can be orderly packed, or hierarchically folded into 46 rod-shaped *chromosomes* (for man), each $\lesssim 10$ μm long, via coiling and winding onto (positively charged) proteins made of histones to form nucleosomes, then coiling again to form a solenoid, and finally via chromatine loops. Imagine the full genetic code, of macroscopic length, packed into a few dozen mesoscopic rods which fit into the nucleus of every cell!

Because of the unique pairing of the four bases, a DNA can be unzipped into two strands, each storing the full information. When a DNA is transcribed in vivo, each sugar S := deoxyribose is replaced by a ribose S', in order to be distinguishable from the original, and each thymine replaced by a uracil. The *genetic code* stored in a thus-obtained, single-stranded RNA is such that each 3 successive rungs form a letter, composed of arbitrary combinations of A,C,G, and U so that there are $4^3 = 64$ letters. Of these, AUG stands for a beginning, and UAA, UAG, and UGA stand for an end. The remaining 60 letters code for 20 different known *amino acids*. Every DNA is composed of a long sequence of subunits, a small percentage of them called *genes*, each $\gtrsim 10^3$ basepairs long, which code for proteins. *Proteins* are chains of $\gtrsim 10^{2.5}$ amino acids; they are the worker bees of a cell. In this way, a DNA stores the information of a long sequence of different amino acids, some 10^6 for man, and is (even) able to synthesize them. It is not clear at this time what percentage of the remaining DNA segments – among them transposons and retroviral sequences – serves exclusively their host, or the host's future evolution, or perhaps nothing useful at all (junk DNA).

Among the basic organic molecules are also the *chlorophylls* which use photon power during the *photosynthesis*of plants to drag electrons to one side of the thylakoid membrane, ready to attract an equal number of protons from the other side whose electrostatic energy (and free-fall momentum through a channel) is used for ATP synthesis. Chlorophyll is related to *haemoglobin*: replace the central magnesium atom in its porphyrin head by an iron atom, and you get one of the four oxygen- and CO_2-binding haem groups in the haemoglobin of the red blood corpuscles of animals. The chemistry of life shows remarkable uniformity and order!

Essential for life are also the pumps which pressurize the blood of animals to make it *circulate*, or the water in plant roots to make it rise, their *hearts*. Remarkably, the human heart has the longest lifetime measured by the number of its beats, $10^{9.6}$, some four-times more than the typical 10^9 beats of animal hearts. The analogous situation for plants is ill-treated in the literature [Kundt, 1998c]: Large trees can lift a ton of water per day to their crowns, part of it overnight, and occasionally under conditions of saturated water-vapour pressure. They can hold their water columns by means of capillarity and osmotic suction which replace evaporation losses. The motor driving the rise of water is generally transpiration. But there is no car without a starter: The roots of plants take in water from the ground via osmotic suction, and a *reverse osmosis* is required in order to allow for a second osmotic pull of their crowns. This faculty has been known for 275 years through the phenomenon of *exudation*, or root pressure, which can reach 6 bar in tomato plants. Root pressure is provided by millions of mono-cellular mechanical pumps in the endodermis and pericycle of the root-hair zone, in all young root tips, $\lesssim$ weeks old, which use some thousand mesoscopic filter valves each – *plasmodesmata* – to achieve the indispensable reverse osmosis.

Another essential for life is *metabolism*. Among the best examples of metabolism is the South-American bull frog, 20 cm long, with a splendid appetite, which eats everything up to the size of 1.5-m long snakes which in turn do not mind eating frogs. It is a matter of speed and strength whose head enters the other's throat for good. Once its head has been swallowed, the victim's further fate is left exclusively to subconscious processes which squeeze it down the gullet and apply all sorts of organic and inorganic chemistry, involving rough-surface wall catalysis in the stomach, to dissolve and metamorphose it into the other's replenishment. It works both ways, and takes a day or longer. Most impressive is the chemical perfection at which complex creatures are reprocessed, with seemingly quite modest growth of the overall entropy.

Even *plants* can digest animals, the meat-eating ones. Moreover, plants use the atmospheric carbon available in the form of carbon dioxide, via *photosynthesis*, using chlorophyll to convert daylight into electric voltage which makes endergonic carbon reactions feasible. These photovoltaic engines are carefully constructed in maximising integrated sunlight during sparse supply, via suitable orientation and chemistry, and by avoiding burnout during excess supply, through non-photochemical quenching (dissipation) via zeaxanthin and possibly lutein. Animals owe their existence to carbon-providing plants.

14.4 Mobility and Senses

Not only animals *move*: *plants* can do so to attach themselves to support, to maximise photosynthesis, and to catch insects; time-lapse movies show this impressively. Whereas geotropy and phototropy are steered chemically, fast motions, like the collapse of mimosa leaves under touch, use joints activated by electric ion pumps. Plants can even enhance their metabolism 10^2-fold, agitated by HCl, like the voodoo lily during blossom, and thereby raise their temperature transiently by $\lesssim 22$ K in order to volatise its odour.

Returning to animals, is it not impressive how brilliantly they manœuvre? Think of monkey jumps, bridging 10-m separations between the crowns of tall trees. Think of the cliff-leaper antelope's jumping $\lesssim 8$ m vertically from rest when threatened, sword-fishes (or wahoos) shooting through the water at $\lesssim 75$ km/h, cheetahs racing at $\lesssim 105$ km/h, certain birds {common swifts, Stachelschwanzsegler} reaching {144, 335}km/h, or dolphins jumping $\lesssim 6$ m vertically out of a modest-sized water basin! Or think of flies landing head over heels on the ceiling. Long-jump records list 14.3 m for a certain stag (Weißwedelhirsch), or 13 m for a giant kangaroo. The minimum acceleration, evaluated for vertical jumps in units of the terrestrial g multiplied by the ratio of height over runway, reaches 200 for fleas and even 400 for a certain beetle! (Man does not take easily more than $6\,g$). A lot more data of this kind can be found in [McMahon and Bonner, 1983].

All these phantastic dynamic achievements by animals can apparently be traced back to actin-myosin ratchets in their *muscles*, shortening them stepwise by small multiples of an angstrœm, each step being powered by an *ATP* molecule. Even more impressive than the cooperative strength of their muscles is their control, often requiring complex steering of multiple limbs, in reaction to rapidly changing external conditions. We thus arrive at the miracle of the senses in animals without which there would be no controlled actions.

School wisdom talks of man's five *senses*: seeing, hearing, feeling, smelling, and tasting, with perhaps orientating added as a sixth sense. But let us take a closer look: Seeing usually refers to mapping the surroundings electromagnetically in the visible frequency band. Rods and cones on the human retina are sensitive to total intensity and three separate frequency intervals respectively, allowing for coloured seeing. Two overlapping eyesights allow for stereoscopic sharpening of distance estimates. Bees and butterflies are sensitive at UV frequencies. Bees can also perceive the direction of linear polarization (of scattered sunlight, revealing the solar position on cloud-covered days). Certain birds have recently been found to be sensitive to $\lesssim 5$ different colours. Many animals are sensitive to IR frequencies, among them snakes, lice, mosquitos, and certain beetles which feed on burnt wood. Perhaps the night-time hunters can likewise see at far-IR (thermal) frequencies. When qualitatively different sensors among different animals are counted separately, *seeing* alone involves at least 10 senses.

Hearing involves another ≥ 6 senses: Consider the chosen frequency ranges, ordinary sound between 16 Hz and 20 kHz for man but *ultrasound* extending up to 0.3 MHz for certain bats, butterflies, dolphins, and also for fish, which sense vibrations in the water with their lateral-line organ. Individual tones can be heard to an accuracy of 10^{-3}, and even the phases of repeated pure tones can be perceived. With two separate ears recording loudness ratios, and arrival times down to 10^{-5}s ($\lesssim v_{max}^{-1}$), directions of sources are routinely inferred. Finally, ultrasound is used actively by bats, flying dogs, certain owls, and sea mammals to create *radar* maps, i.e. to acoustically map their surroundings. They do this in various ways, by means of shouting quasi-continuous, mono-frequency signals whose echos are Doppler blueshifted for approaching reflectors, or else by shouting short pulses whose echos arrive intermittently. Clearly, at least 6 different types of organ are involved in all these acoustic facilities.

Feeling has perhaps only 4 degrees of freedom, one for *pressure*, one for *pain*, one for *pleasant temperatures*, and one for *unpleasant* (low or high) *temperatures*. Receivers are distributed not only throughout our skin but also throughout our interior, telling us the relative orientations of all our limbs (of which we are aware even with eyes closed!), and usually warning us in the case of some malfunction. As the orientation of each joint involves up to three angles, controlling our attitude involves controlling well over 10^2 angles.

There are only 5 *tastes* which our tongue can distinguish: sweet, sour, salty, bitter, and umami (produced by mono sodium glutamate), plus a joint fortissimo when we eat *hot* spices. All the other often-called tastes are rather *smells*, sensed by our nose. Smells are based on distinct molecules reaching one of the receivers; we are sensitive to over 10^4 of them. More than 10^{-2} of the genetic information codes for this sense. Smell can be important for animals to find food, find the partner, or find home. Camels can even smell water vapour, and bees can in addition smell CO_2 (to avoid being poisoned whilst asleep).

Animals can sense their *orientation* w.r.t. gravity, and their *accelerations* in two perpendicular directions, three vital senses for mobility; their accelerometers tend to be located in their ears. In addition, flies, bees, and flying beetles avail themselves of tachometers, viz. fans whose bending records their speed relative to the ambient air so that they can reach their goal even in bad weather. Again in addition, one pair of wings in flies has converted to gyro compasses which guide their audacious overhead manœuvers, bringing the number of orientation senses up to ≥ 5.

Certain species of fish and mammals in deep waters, and birds travelling around the globe require at least one additional sense for long-distance navigation. They are thought to sense the electric Lorentz force $e \mid \boldsymbol{\beta} \times \boldsymbol{B} \mid$ when moving at a known velocity $c\boldsymbol{\beta}$ across the terrestrial magnetic field $\boldsymbol{B}$, with sensitivities of their voltmeters reaching down to 5 nV/cm. In homing doves, the Lorentz-force meters may be located in the upper parts of their beaks where nerves with interspaced magnetite crystals have been found in 1997.

In reviewing the *external senses* of animals, I have already covered ultrasound as a means for radar orientation of several flying nocturnal predators. Another *active mapping* is used by fish in turbulent, turbid waters, in the form of self-generated *electric* signals whose subsequent field configuration maps the electric conductivities of their surroundings. Such fish strike the eye by having a straight, stiff body. Its signals have voltages of $\lesssim 10$ V, and either pulse frequencies between 2 Hz and 1.6 kHz, or a continuous oscillation at $10^{2.5\pm0.5}$Hz whose precise frequency can be tuned, in reaction to rivals. Other fish, in the deep sea, use the cold light emitted by certain symbiotic bacteria which generate the photons with ATP-powered lamps, combining luciferin with luciferase at a high light-harvesting efficiency.

Besides the external senses, every animal enjoys an almost comparable number of *internal senses* which communicate, so to speak, its house-keeping data. Among them, *orientation* of all its movable parts has already been mentioned. *Breathing* is enforced by an increase in the percentage of CO_2 in the blood, *thirst* by the blood's concentration, and *hunger* by the fullness of the stomach plus the difference in sugar concentration between arterial and venous blood, modulated by air temperature. *Appetites* steer the versatility of the food and its fermentation in the gullet. Corresponding senses notify an individual of the degree of fullness of its guts or bladder, or preference for express *clearance* by sneezing, coughing, spitting, or vomiting. Web spiders avail themselves of a multitude of specialised glands for the production – and subsequent extrusion – of different silks. And a creature feels *tired* when its brain wants to re-configure its software. *Sleep* is guarded: actions planned during a dream tend to be held back before being issued. Finally, there are the *sexual* demands.

In addition to the conscious internal senses, there are the *subconscious* ones which act in warm-blooded animals as reliable *feedback circuits*: The blood *temperature* tends to be stabilized at the 10^{-3} level whereby individual thermometers sense fluctuations of only 10^{-2}K, and where the key thermostat sits in the front part of the hypothalamus, controlling the temperature of the main arteries supplying the brain. Similarly, the pH *value* of the blood is buffered (via CO_2) near 7, at the $10^{-2.5}$ level. And so are the blood *pressure*, its CO_2 contents, its contents of *sugars*, its *immune* system, and the blood's non-coagulation even when veins are wounded. Water-living animals solve the further problem of maintaining a jump in osmotic pressure of their body fluid to the ambient medium, upward or downward, depending on its salinity. Animals are well-controlled engines working at high reliability, steered by thousands of senses.

Every sense requires detectors, conductors, memories, and evaluation, plus a broadcaster in the case of active mapping. Mechanical *detectors*, for example, use 'hair cells': special sensorial cells topped by a skew conical bunch of some 50 plus 1 special (peripheral) tiny mobile hairs, protrusions of the outer cell membrane, whereby the special hair, the kinocilium or *antenna*, has a balloon-like head and contains two central microtubules surrounded by a ring of nine pairs of mobile microtubules. This quite universal array succeeds in converting minute mechanical excitations reliably into electric signals.

All signals inside animals are transferred via *nerve cells (neurons)*, or rather via their ($\lesssim$ body-long) *axon* appendages, at speeds of $\lesssim 10^2$m/s. As outcrop of a cell, an axon is a narrow tube, $10^{0\pm1}\mu$m wide, formed from the poorly insulating cell membrane and from the electrolytic cell fluid in its interior which makes it a poor conductor. Whenever necessary for insulation, axons are encased in thick myelin sheaths, regularly interrupted by Ranvier gaps (or *nodes*, for signal amplification). When an axon fires, a standard electric pulse propagates along it, all the way to its end, amplified at 1.5-mm separations by Na^+-K^+ pumps in the membrane. About 10^6 such axons connect every eye with the brain. The ends of axons branch out, like the neuron bodies themselves (into dendrites), and make connections with other neurons, muscles, or glands, via synapses which establish one-way chemical connections.

Animal *brains* are *neural networks*, consisting of huge numbers of neurons whose states can be influenced by 10^3 to 10^4 input signals each, via *synapses* of different strength. The new state of a neuron determines its output which is subsequently transferred to other neurons. The human brain has $10^{10.6}$ neurons, and $10^{14.3}$ synapses, compared to $\{10^{3.5}, 10^{8.5}, 10^{9.5}\}$ synapses of a {worm, fly, bee}. Whereas human neurons fire at intervals of $\lesssim 10$ ms, digital computers approach the ns time level. Nevertheless, biological neural networks can be vastly superior to our best digital computers in applications like face recognition, speech, mobility, and other mental achievements, by processing vastly in parallel. Of course, they require learning. Concerning speed, a typical robot needs some 10^3 connections per second, speech some 10^6/s, image processing $\gtrsim 10^{10}$/s, and cognitive properties $\gtrsim 10^{14}$/s; man uses 10^{16}/s. Man's $10^{14.3}$ synapses are thought to realise what philosophers classify as instinct, intellect, mind, and soul.

There is no brain without a memory. Accidents and experiments have shown that man has at least 3 *memories*: a short-time memory saving fresh news electrically for $\lesssim 0.7$ s, a transient chemical memory storing its contents for $\lesssim$ an hour, and a lifetime chemical memory which has a practically unlimited storage capacity. During serious accidents, some epoch preceding it tends to remain unrecorded, of order $\lesssim$ h; but even a loss of two days' memory – before getting cooled down by a river (to 13.7°C) – has been reported, suggesting yet more complicated data handling.

The preceding paragraphs were thought to convey, or re-emphasise the impression that life obeys the laws of physics rigorously, without exception. Does it? Among the gaps in our understanding are how a plant, or an animal is assembled according to its DNA blueprint in a fully *self-organised* way, without a commander. How does each of the $\gtrsim 10^{16}$ forming cells of a large organism know its particular part in the grand plan? Via the concentration ratios of a number of signalling proteins? It has been found that when a female germ cell, an egg, has divided into two halves which are carefully separated, or even into four (equal) parts after the second division of cells, independent and quite normal animals can be grown from the four partial cells, but that in general, such a procedure is no longer successful after the third division: cells start to specialise [Sergeev, 1978]. A Nobel prize has been assigned for showing how the proper cell specialisation can be accomplished through concentration gradients in a forming individual. But why is self-organisation quasi-infallible, more than 10^{16} times, for some 10^2 different cell types, steered solely by the DNA and the local environment of a forming cell? How does a larva metamorphose into a butterfly via a pupa? How does the flour beetle even remember what its larva had learned (viz. to turn only to the right when moving in a labyrinth)? For me as a non-specialist, this apparent infallibility of life is an enigma. An enigma that, nevertheless, has not given any hint of a violation of the laws of physics.

14.5 Evolution

How has life on Earth reached the fantastic complexity, and adaptation sketched in the preceding section? What was the *origin of life*? Fossils tell us that *surface life* on Earth *began* at 3.4 Gyr in the past, after the cosmic bombardment had declined to a tolerable level and after an uncertain initial CO_2 atmosphere had given way to a CH_4-dominated atmosphere with a rising N_2 and O_2 contribution, via outgassing and H losses through the exosphere. Eucaryotes entered the stage at -1.4 Gyr, and oxygen-breathing multicellular organisms at -0.7 Gyr, profiting from the 14-times higher available energy compared with anaerobic metamorphism. When UV-screening by atmospheric O_3 became effective, at -0.4 Gyr, the continents were conquered by life, and a roughly exponential, or rather piecewise logistic *diversification* of species set in – interrupted by some ten mass extinctions (due to climatic perturbations caused by giant ($\gtrsim 10^{18}g$) meteoritic impacts?) – to a present number of $\lesssim 10^{7.7}$ (non-interbreeding) species, or $\lesssim 10^4$ families, with hominids entering the stage at (only) $-10^{-2.5}$Gyr, and humans at $-10^{-3.8}$Gyr, according to the mitochondrial DNA clock. Note that the exact *tree of life* – the ordering of plants and animals as they evolved from each other – is just being explored, with still major discrepancies between the results of the *morphological* (fossil) and *molecular* (DNA) method [Nature *406*, 230; *408*, 652 (2000)].

In the process of *diversification*, or *evolution*, minor adaptations occur quasi-continuously, from generation to generation, whereas major mutations have happened unresolvably fast, probably within one generation, because populations become extinct if represented by too few members ($<10^2$). How do mutations come about? Based on the work by Erwin Schrödinger, John von Neumann, Manfred Eigen, Leslie Orgel, Stanley Lloyd Miller, and many others, Freeman Dyson [1985] distinguishes between biological hardware and software: *Hardware* consists of proteins (composed of 20 amino acids), is stable in a reducing atmosphere, is responsible for metabolism, serves as a host, and can (nowadays) synthesize ATP. *Software* consists of nucleic acids, is unstable outside of a cell (to hydrolysis), is responsible for replication, first appeared as a parasite, and has the various forms of RNA as its representatives. Dyson argues that life started with cells; enzymes came second, and genes much later. Life started as hardware performing metabolism and reproducing without being able to (identically) replicate, and was only much later conquered by the DNA which subdued an (archaean?) cell, and turned it into a eucaryotic cell, with guaranteed replicability. This *double origin of life* would have taken the 2 Gyr before the emergence of eucaryotes. Did it take place underground, in the much more sheltered rocky crust of Earth, yielding bacteria which could feed on buoyantly rising (abiogenic) carbohydrates [Gold, 1999]? Does the fact that all biological DNA is *right-handed* (rather than occurring as a mixture of right- and left-handed helices) tell us that replicative life on Earth had a unique beginning, i.e. that after the first endosymbiosis of a cell with a DNA, replication was so fast that a second creation could not compete?

Once we have an idea of how life may have come into existence, we are back to the problem of *evolution*: how stable is replication? RNA copying in vitro shows an error probability of 10^{-2}, but enzymes reduce this relative error rate in viruses to 10^{-4}. The two-fold redundancy of the DNA (over RNA) plus proof-reading then roughly square this unreliability (to $10^{-8\pm2}$) and make faithful replication possible. Still, all biological evolution apparently owes its existence to the rare changes in DNA transmission which take place via sexual reproduction [Forsyth, 1986], i.e via changes in Dyson's software. Life strives to reproduce sexually, both plants and animals. *Reproductive instability* is the apparent reason for the evolutionary conquering of niches. Almost all higher organisms can reproduce sexually, though many of them can in addition multiply asexually, i.e. produce *clones*. Plants do so quite regularly. Cloning appears to be rather stable, as has been shown in a 22-yr experiment with $10^{4.13}$ generations on an infusorian slipper animalcule, a unicellular creature which divided twice per day on average and was permanently prevented from meeting a partner.

How does the DNA evolve? Random exchanges of DNA basepairs do not qualify: almost all of them are lethal. Viable DNA molecules should be predetermined. Layzer [1990] talks of β-genes on the DNA which control *genetic recombination* during reproduction, as opposed to the regulatory α-genes which exclusively serve the individual which carries it; $\lesssim 90\%$ of the DNA are thought to encode strategies for evolution. And even with such an elaborate effort of nature to optimise its creatures, it is difficult to see how a macrostep of evolution – involving a change of the habitat by a specialised creature – could have been taken essentially within a single generation, say via gene splicing (exon shuffling). Gold [1999] talks of '*Darwin's dilemma*'; he reminds us of the omni-presence of symbiotic phenomena, and speculates about an intervention of micro-organisms, perhaps something like a virus infection affecting reproduction, simply because rare phenomena require large numbers of trials which have not been available at the multicellular level. Note that a macrostep may require dozens of identical mutations within one generation, in order not to die out again. How, for example, did sperm whales come into existence, with their more than unusual constitution; via sea elephants?

Time and again, *biodiversity* and its impressive stability (despite all manmade hardships) make me wonder whether life on Earth can really be considered a consequence of the laws of physics. A case in favour may be the empirical law of biogenetics stating that the *ontogenesis* is a shortened repetition of the *phylogenesis,* i.e. that certain preferred construction pathways have been pursued right from the beginning of life, and continued to ever higher levels of organization. The structural hierarchy is also a temporal hierarchy; evolution has followed a scheme which stood the test of stability at every level of sophistication.

But there is also a case against the straightforward lawfulness of life: *mimickry.* Certain poisonous animals warn their potential predators by provoking colours: snakes, frogs, beetles. For almost all of them, there is a rather identical-looking creature which lacks the property of being poisonous. Did the poisonous species exist first, and is genetically pursued by the harmless species, via evolution? Or are we lacking some higher insight? Have the laws of physics been devised such that life is a cosmic imperative, with all its wonders?

14.6 Anthropic Principle

The preceding section has led us to the theme of the present one: are the laws of physics a consequence of our existence? Would a minor change of them forbid the existence of Homo Sapiens? In this most general of its versions, the *anthropic principle* is no more than a provoking hypothesis. Yet for us to exist, there are many seemingly arbitrary properties of our cosmic environment which had better not be different from what they are. For instance, the mean *mass density* in the Universe must not be much larger than the critical density (1.21), or else the Universe would have recollapsed before life could have evolved. The biological evolution has taken all the time of a habitable Earth, i.e. the radioactive *age* of Earth is not distinctly larger than its biological (genetic) age. This age, in turn, agrees with a significant fraction of the Sun's main-sequence age: The *Sun* must not be more *massive* than it is, by a factor of order two, or else it would not shine steadily for long enough. On the other hand, if the Sun were half as massive as it is, its spectrum would be too soft for certain chemical reactions to be energised by its light, (much redder than its peak at yellow-to-orange frequencies). It must be an early G star. Also, the mass density in the Milky Way, and hence the ambient stellar density must not be larger by an order of magnitude, or else the Solar System would probably have tidally interacted with one of its neighbouring stars to an extent that had intolerably enhanced the Earth's eccentricity and hence yearly temperature variation.

If *water* is an essential for life – and Sect. 14.2 has listed 12 supporting reasons – a habitable planet, or moon, must have the proper distance from its (heating) star in order to have a surface temperature within the liquid range for water. An absorbed solar energy flux $\alpha\pi r^2 S_\odot$ which is reradiated isotropically at a mean *temperature* T as $\epsilon 4\pi r^2 \sigma_{SB} T^4$, where α and ϵ are the mean (visible) absorptivity and mean (thermal) emissivity respectively, gives rise to an average T at a distance r from the Sun of

$$T = [(\alpha/\epsilon)(S_\odot/4\sigma_{SB})]^{1/4} = 278 \text{ K } (AU/r)^{1/2}(\alpha/\epsilon)^{1/4} \ . \tag{14.1}$$

For present Earth, effective values of $\alpha \approx 0.7$ and $\epsilon \approx 0.6$ yield $T = 289$ K $= 16°$C. More exactly, Earth radiates with an $\epsilon \approx 1$ from above the troposphere, at an effective temperature near 250 K. Of course, α drops with an increasing percentage of ice and snow cover, and ϵ depends sensitively on the heat transfer through the troposphere, with the result that a once totally snow-covered Earth is thought to never warm up again, i.e. to suffer runaway glaciation, and a once ice-free Earth is thought to suffer a runaway greenhouse death. The fact that Earth still finds itself midway between these thermal Scylla and Charybdis, after some $10^{9.7}$yr of slowly rising solar input and strongly changing atmospheric conditions, should be attributed to its suitable distance from the Sun which is thereby fixed within some 10% of its a priori range. Mars is far too distant for liquid surface water, and Venus is far too near; Earth's AU separation is just right, see Rampino and Caldeira [1994].

One can now ask the question of how many *habitated planets*, or *moons* there may exist in our Galaxy. An answer is given by the *Greenbank*, or *Drake formula* which estimates this number N_{life}:

$$N_{life} = N_{G\text{-}stars}\ p_{separ}\ p_{envir}\ p_{spin}\ p_{chem}\ p_{life} = 10^{4\pm4} . \qquad (14.2)$$

Here $N_{G\text{-}stars} = 10^{10}$ is the number of Galactic G stars, $p_{separ} :=$ probability of having a satellite at suitable separation ($\approx 10^{-1}$, as concluded above); the environmental probability p_{envir} considers potential hazards to life such as a nearby companion, a collision with an interstellar cloud, a destructive late meteoritic impact, a very near SN ($\lesssim$ pc), insignificant magnetospheric and atmospheric screening, or any overlooked additional hazards, and judges their union optimistically as $\lesssim 0.3$, and the final three probabilities concern a sufficiently short spin period (for moderate temperature variations; of $\lesssim 4$ d), a suitable chemistry of soil and atmosphere (with the necessary $\lesssim 40$ chemical elements, which will depend on the satellite's size, and may be controlled by deep-rooted volcanism including plate tectonics, and by the tidal influence of a nearby moon), and whether or not life has originated on this satellite without intolerable delay; their product has been optimistically set $\lesssim 0.3$. The existence of our home planet guarantees $N_{life} \geq 1$. In words: there may be between 10^8 and 1 habitated satellites in our Galaxy, with 10^4 as an educated guess. When one narrows in this estimate to the presence of a *civilization* – whose lifetime may be limited to some 10^3yr, due to rapid self-destruction – the answer may read $N_{civil} = 10^{1\pm1}$. Civilized life is likely to be a rare phenomenon in the Universe.

A number of further constraints on the presence of life have been derived in the past century, see Press and Lightman [1983], and Barrow and Tipler [1986]. Among them are an estimate of the *mass ratio* m_e/m_p of electron and proton which should be $\ll 2$ for atomic nuclei to be bound more tightly ($|E_b| \gtrsim \alpha m_p c^2$) than atoms ($|E_b| \approx Ry := \alpha m_e c^2/2$), $\alpha :=$ fine-structure constant, but also $\gg \alpha^2$ for atomic nuclei to be in general β-stable, a fairly narrow range which is also dictated by the existence of planets, see Problem 1.3.6. Further, the constraint that the solar surface spectrum be suitable for photosynthesis, $h\nu_\odot \approx Ry$, implies

$$\alpha_G \approx \alpha^{12}(m_e/m_p)^4 = 10^{-38.7} \qquad (14.3)$$

for the *gravitational finestructure constant* $\alpha_G := Gm_p^2/\hbar c$, remarkably close to its value $10^{-38.23}$. A different estimate, $\alpha_G \approx \alpha(m_e/m_p)^{1/2}(a/l)^4$, follows from the mechanical stability of creatures of size $\lesssim l$, where $a := \hbar^2/e^2 m_e = 10^{-8.28}$ cm is the Bohr radius which determines atomic separations.

Next, the *nuclear abundances* cooked inside stars depend sensitively on the binding energies of certain nuclei, to within 10^{-2} or less, in particular of Be^8, C^{12}, O^{16}, and of the (unbound) di-proton. Correspondingly, *molecular binding* energies must be fine-tuned in order to allow for the existence and properties of e.g. the DNA. *Replicability* in biology depends on the identity, and stability of DNA and other molecules, which are guaranteed by quantum mechanics. Without *weak interactions*, there would be no (long-lived) main-sequence stage of (G-) stars. The anthropic role of *neutron stars* is not (yet) established; chaps. 9 and 10 discuss their possible significance as Galactic particle boosters to extreme energies whose (cosmic-ray) spallation products contribute significantly to the light elements Li, Be, and B, and whose impacting on the terrestrial atmosphere creates the cosmic air showers.

Are any of these apparent constraints dispensible for life? Does the existence of Homo sapiens on Earth really tie down the *laws of physics* to what they are, or at least tie down its dimensionless constants? There are yet too many unanswered problems, in particular in biophysics, to make ultimate statements. But to me, the wonder of biological self-organization and evolution asks for large sets of very special molecules whose unique properties allowed for a hierarchical growth of structure right from the beginning, thereby defying a rapid approach of local thermal equilibria. The anthropic principle may be more than a challenging hypothesis; see Plate 14.

15. Answers to Problems

1.2.1: $P = 10^{2.6}$s $a_{10}^{3/2}(2M_\odot/M)^{1/2}$, $P_{\min} = \{10^{4.4}, 10^{1.4}, 10^{-3.1}\}$s $= \{7$ h, 0.4 min, 0.8 ms$\}$.

1.2.2: $R = 10^{10.9}$ cm$(M/M_\odot)^{1/2}\Delta t_{(10)}^{-1/2} \approx R_\odot$.

1.3.1: $t = \{12$ d, $\{^{8\,\mathrm{M}}_{0.6\mathrm{k}}\}$yr, $\lesssim 3$ Gyr$\}$.

1.3.2: $P/\mathrm{yr} = \{10^{7.9}, 10^{8.1}\sqrt{z/\mathrm{kpc}}\}$.

1.3.3: $E = 10^{18.3}$eV $B_{-5.2}$.

1.3.4: $e^{-\tau} = \{1 - 10^{-4.2}, 1 - 10^{-2.7}, 10^{-4.3}\}$.

1.3.5: $\Delta t_g/\Delta t_h = 10^{-3.1}\sqrt{M/M_\odot}$.

1.3.6: a) $n_{elm} \approx (\pi\alpha/\lambda_e)^3 = 10^{24}$cm^{-3} $(=$ overestimate$)$, $n_{grav} \approx (\pi\alpha_G/\lambda_e)^3$ $(M/m_N)^2 = 10^{18}$cm$^{-3}M_{28}^2$ $(=$ gross underestimate$)$, $n_{\max} = (2.5/\lambda)^3$ $= \{10^{30}, 10^{39.8}\}$cm^{-3}. b) $M_{Ch} \lesssim m_N \, \alpha_G^{-3/2} = 10^{33.5}$g , $M_F = m_N$ $(\alpha/\alpha_G)^{3/2} = \alpha^{3/2}M_{Ch} = 10^{-3.2}M_{Ch}$, $M_{BH} = M_{Ch}/\sqrt{n(\lambda_N/2\pi)^3} \gtrsim M_{Ch}$.

1.5.1: $\Delta\lambda/\mathrm{\AA} = \{10^{-0.8}, 10^{0.2}\}$.

1.5.2: $d/\mathrm{Mpc} = \{1.4, 14\}$.

2.1.1: a) $B \lesssim \{3\,\mu\mathrm{G}, 10$ kG$\}$, b) $B \lesssim 10^{5.8}$G $n_{18}^{1/2}v_8 \lesssim$ MG.

2.2.1: $\mathcal{M} = \{\sqrt{9/5} = 1.34, \sqrt{80} = 8.9\}$, $\rho_+/\rho_- = \{1.5, 4\}$.

2.3.1: $r \leq 10^{19.4}$cm $(\dot{M}_{(-6)}v_8/p_{-12})^{1/2}$, $t \lesssim$ Myr.

2.3.2: $\Delta r/r \approx 1/12$.

2.4.1: $r = 10^{19.3}$cm $(L_{37}/n_2)^{1/2}$, $t = 10^{5.7}$yr.

2.7.1: $r = 10^{16.2}$cm $(\Delta M_{(0.5)}\,\sigma_{-24}\,m_p/m)^{1/2}$, $t \lesssim 10^{7.5}$s $\Delta M_{(0.5)}(\sigma_{-24}m_p/mE_{51})^{1/2}$.

3.1.1: a) $\Delta W/W \leq 10^{-15.5}(eE)_{-8.8}$, b) $\Delta W/W = 10^{-2.6}\,\gamma_4^4 \, / \, a_4^2\,(eE)_{-8.8}$.

3.1.2: $\tau = \{10^{0.1}$ms, 10^2yr, $10^{6.1}$yr$\}$.

3.1.3: $\gamma_{slc} = 10^{13.9}B(r_*)_{12}\,m_e/m$.

3.2.1: a) $T \gtrsim 10^{3.1}$K , b) $\gamma \lesssim 10^{3.6}/\sqrt{B_0}$, c) $\gamma \lesssim 10^{0.25} = 1.7$.

3.2.2: $T_r \gtrsim 10^{7.1}$K $r_{13.2}^{-2}$.

3.2.3: $f = \gamma_{slc}$.

3.3.1: $n\lambda^3 \approx 1.2\,\pi/4 \approx 1$.

3.3.2: $\langle E \rangle = \{E_u/2, E_u(E_l/E_u)^{g-1}(g-1)/(2-g), E_l\}$.

3.3.3: $\cos\vartheta = \nu_{5.5}^{-2}$.

3.3.4: $R = $ AU $Z_{-5}\,\dot{M}_{(-6)} \, / \, v_8$.

3.3.5: $\tau = \{(T_2/T_1)^4, (T_2/T_1)\}$.

4.1.1: $s/nk = \{18, 11, \gtrsim 88, \lesssim 2.5, 2\pi^4/45\,\zeta(3) = 3.6\}$.

4.1.2: $T_F/\text{K} = \{10^{4.9} n_{23}^{2/3}, \gtrsim \{^{10^{9.5}}_{10^{6.2}}\}, \approx 10^{12.3}\}$.

4.1.3: $t = \{5 \text{ h } R_1 T_{2.7}^{-3}, 17 \text{ min}/n_{10}, 10^7 \text{yr}\}$.

5.1.1: $\Omega \gtrsim \{10^{-3.3} \text{ s}^{-1}/ r_{11}^{3/2} B_4^{1/2}, 10^{-6} \text{s}^{-1}/r_{11}^2 B_4\}$ for $m = \{3, 2\}$.

5.2.1: $p/\text{bar} = 10^{-5.4} T_{2.5}^4 / Z^2$.

5.3.1: $t_{dec} = \{10^{-0.8} \text{s } x_{0.7}^2, 10^{7.4} \text{yr } \sigma_{14}, 10^{1.4} \text{yr}, 10^{1.7} \text{yr } \nu_{12}/A_{14}(\delta\rho/\rho)_{-4}\}$.

6.1.1: $t/\text{yr} \lesssim \{10^{10.5}, 10^{-0.5} r_{11}^{3/2}\}$.

6.1.2: $10^{14.4} \lesssim \nu/\text{Hz} \lesssim 10^{15.9}$.

7.1.1: $T_{crit}/\text{K} = \{10^{0.8}, 10^2, 10^5\}$.

9.1.1: $E_{th} / | E_{bdg} | = \{1/2, 1, \infty\}$.

References

Bahcall, J.N. and Ostriker, J.P., 1997: *Unsolved Problems in Astrophysics*, Princeton Univ. Press.

Barrow, J.D. and Tipler, F.J., 1986: *The Anthropic Cosmological Principle*, Oxford University Press.

Begelman, M.C., Blandford, R.D. and Rees, M.J., 1984: Theory of Extragalactic Radio Sources, *Rev. Mod. Phys.* **56**, 255–351; in particular its Appendices.

Burrows, A., 2000: Supernova explosions in the Universe, *Nature* **403**, 727–733.

Crusius, A. and Schlickeiser, R., 1988: Synchrotron radiation in a thermal plasma with large-scale random magnetic fields, *Astron. Astrophys.* **196**, 327–337.

Dröscher, V.B., 1991 (1975): *Magie der Sinne im Tierreich*, dtv, München.

Duve, Ch. de, 1994: *Ursprung des Lebens*, Spektrum Verlag.

Dyson, F., 1985: *Origins of Life*, Cambridge Univ. Press.

Forsyth, A., 1986: *A Natural History of Sex*; German edition [1987]: *Die Sexualität in der Natur*, dtv, München.

Gold, T., 1999: *The Deep Hot Biosphere*, Springer.

Hasselmann, K., 1998: The Metron Model: Towards a unified deterministic theory of fields and particles, in: *Understanding Physics,* ed. A.K. Richter, Copernicus Gesellschaft, pp. 155–186.

Heusler, M., 1996: *Black Hole Uniqueness Theorems*, Cambridge Lecture Notes in Physics **6.**

Hoyle, F., 1975: *Astronomy and Cosmology*, a Modern Course, Freeman and Co.

Hoyle, F., Burbidge, G. and Narlikar, J.V. [2000]: *A different approach to Cosmology*, Cambridge Univ. Press.

Karttunen, H., Kröger, P., Oja, H., Poutanen, M. and Donner, K.J., 2000: *Fundamental Astronomy*, Springer.

Kippenhahn, R. and Weigert, A., 1990: *Stellar Structure and Evolution*, Springer.

Kronberg, P.P., 1994: Extragalctic Magnetic Fields, *Rep. Progr. Phys.* **57**, 325–382.

Krotscheck, E. and Kundt, W., 1978: Causality Criteria, *Commun. Math. Phys.* **60**, 171–180.

Kundt, W., 1972: Global Theory of Spacetime, in: *Proceedings of the 13th Biennial Seminar of the Canadian Mathematical Congress*, ed. J.R. Vanstone, Montreal, pp. 93–133.

Kundt, W., 1990a: ed. of *Neutron Stars and their Birth Events*, NATO ASI **C 300**, Kluwer Acad. Publ.

Kundt, W., 1990b: The Galactic Center. *Astrophys. and Space Science* **172**, 109–134.

Kundt, W., 1996: ed.: *Jets from Stars and Galactic Nuclei*. Lecture Notes in Physics **471**, Springer.

Kundt, W., 1997: Structure of the Galactic Halo and Disk, in: *The Physics of Galactic Halos*, eds. H. Lesch, R.J. Dettmar, U. Mebold, and R. Schlickeiser, Akademie Verlag, 255–259.

Kundt, W., 1998a: Astrophysics of Neutron Stars, Facts and Fiction about their Formation and Functioning, *Fundamentals of Cosmic Physics* **20**, 1–119.

Kundt, W., 1998b: The Gold Effect: Odyssey of Scientific Research. in: *Understanding Physics*, ed. A.K. Richter, Copernicus Gesellschaft, pp. 187–240.

Kundt, W., 1998c: The Hearts of the Plants, *Current Science* **75**, 98–102.

Kundt, W., 2000: Radio Galaxies powered by Burning Disks, in: *Life Cycles of Radio Galaxies*, Baltimore, July 15–17, 1999, eds. J.A. Biretta, C.O'Dea, and P.Leahy, submitted 29 September 1999; or: Jet Formation and Dynamics: Comparison of Quasars and Microquasars, in: *Galactic Relativistic Jet Sources*, Granada, September 11–13, 2000, eds. A.J. Castro-Tirado, J. Greiner and J. Paredes, Kluwer $\gtrsim$ July 2001.

Kundt, W., and Lüttgens, G., 1998: Rings around Planets, Atmospheric Superrotation, and their great Spots, *Astrophysics and Space Science* **257**, 33–47.

Lang, K.R., 1998: *Astrophysical Formulae (I and II)*, Springer-Verlag.

Layzer, D., 1990: *Cosmogenesis – The Growth of Order in the Universe*, Oxford University Press.

Lotova, N.A., 1988: The Solar Wind Transsonic Region, *Solar Physics* **117**, 399–406.

McMahon, T.A. and Bonner, J.T., 1983: *On Size and Life*, Scientific American Books, Inc., New York.

Press, W.H., Lightman, A.P., 1983: Dependence of Macrophysical Phenomena on the values of the Fundamental Constants, *Philos. Trans. R. Soc. London* **A 310**, No. 1512, 323–336.

Rampino, M.R. and Caldeira, K., 1994: The Goldilocks Problem: Climatic Evolution and Long-Term Habitability of Terrestrial Planets, *Annu. Rev. Astron. Astrophys.* **32**, 83–114.

Reid, M. J., 1993: The Distance to the Center of the Galaxy, *Annu. Rev. Astron. Astrophys.* **31**, 345–372.

Reipurth, B., Bally, J., Fesen, R.A. and Devine, D., 1998: Protostellar jets irradiated by massive stars, *Nature* **396**, 343–345.

Reynolds, R.J., 1990: The Power Requirement of the free-electron layer in the Galactic Disk, *Astrophys. J.* **349**, L17–19.

Schneider, P., Ehlers, J. and Falco, E.E., 1992: *Gravitational Lenses*, Springer.

Schwinger, J., Tsai, W.-Y. and Erber, T., 1976: Classical and Quantum Theory of Synergic Synchrotron-Čerenkov Radiation, *Ann. Phys.* **96**, 303–332.

Sergeev, B.F., 1978: *Physiology for Everyone*, Mir Publishers, Moscow.

Spicker, J. and Feitzinger, J.V., 1986: Are there typical corrugation scales in our Galaxy? *Astronomy and Astrophysics* **163**, 43–55.

Trammell, S.R. and Goodrich, R.W., 1996: Hubble Space Telescope and ground-based imaging of the bipolar proto-planetary nebula M1-92: evidence for a collimated outflow, *Astrophys. J.* **468**, L107–110.

Varju, D., 1998: *Mit den Ohren sehen und den Beinen hören – Die spektakulären Sinne der Tiere*, C.H. Beck, München.

Walker, Jearl, 1977: *The Flying Circus of Physics*, John Wiley, New York.

Wu, K.K.S., Lahav, O. and Rees, M.J., 1999: The large-scale smoothness of the Universe, *Nature* **397**, 225–230.

Plates

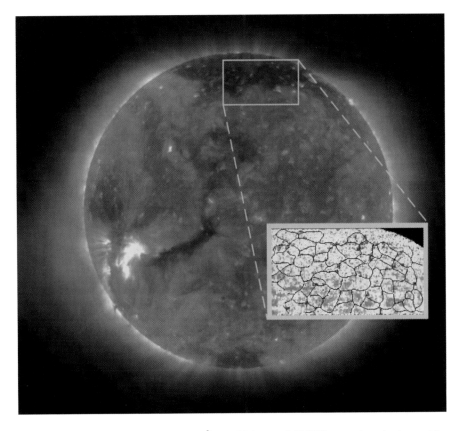

Our Sun in the light of the 195Å EUV line of FeXII – emitted above the chromosphere at 1.5 MK – taken by the SOHO mission in September 1996. Bright regions indicate hot, dense material confined magnetically, while dark regions imply an open magnetic-field topology, known as coronal holes. The inset provides a close-up Doppler-velocity map of same gas in the light of the 770Å EUV line of NeVIII: dark blue represents an outflow (blueshift) at 10 km/s, dark red a corresponding downflow. Apparently, outflow prevails in the boundaries of the 'honey-comb' shaped pattern of the magnetised network. Whilst the Sun looks different at every frequency ν and epoch t, sunspot areas can always be recognized because cooler and warmer subregions tend to closely correlate (in projection onto the surface). [Courtesy of Kenneth R. Lang: The Sun from Space, Springer, Berlin, Heidelberg 2000]

Plate 1

First image by the Hubble space telescope, HST, on 1 April 1995, of the star-forming Eagle Nebula = M16 in Serpens, at a distance of 2.3 kpc. The map is a false-colour superposition of a red, green, and blue photograph taken at the wavelengths {673, 656, 501}nm of {[SII], Hα, [OIII]}. The longest of the 3 'pillars' measures 1 lyr. [Courtesy of Jeff Hester and Paul Scowen (Arizona State University), NASA and STScI]

Plate 2

Inner part of the Egg Nebula, CRL 2688, a young planetary nebula (several 10^2yr) at a distance of 1 kpc, mapped with the HST in red light in 1996. The full nebula has an angular diameter of $25''$. Do we see a twin jet, approaching us in the NW at $\gtrsim 20°$, whilst the innermost part of its antipode is occulted by a disk? Should the circular stripes be understood as density enhancements in the windzone which are crossed by the two jets? The ages t of PNe can be estimated from their radial-expansion speeds, and range through similar values as those of SNRs: $10^{2.5}$ yr $\lesssim t \lesssim 10^{4.5}$yr. [Courtesy of Raghvendra Sahai and John Trauger (JPL), the WFPC2 Science Team and NASA]

Plate 3

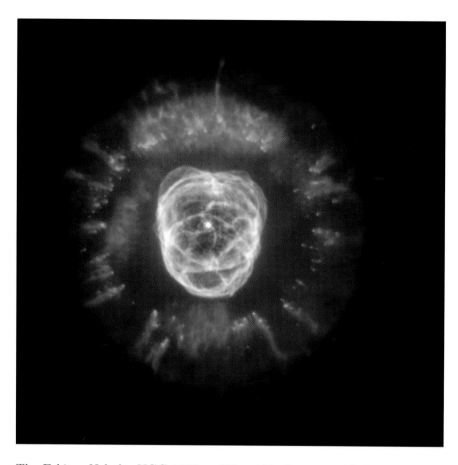

The Eskimo Nebula, NGC 2392, a 10 kyr-old planetary nebula in the constellation Gemini, at a distance of 1.6 kpc, imaged with the HST on 10/11 January 2000. Superposed in {red, green, blue, violet} have been exposures at {6583, 6563, 5007, 4686}Å emitted by {[NII], Hα, [OIII], HeII}. The Eskimo's overall diameter exceeds 1 pc. The comet-like streaks in the outer ring may have been blown by a faster wind overtaking a slower preceding one. [Courtesy of NASA, Andrew Fruchter and the ERO Team (STScI)]

Plate 4

Colour-enhanced photograph of the (inner part of the) Crab Nebula, $d = 2$ kpc: The filamentary network of SN ejecta looks red, due to Hα, whereas blue synchrotron radiation dominates in the nebula's interior. Contrary to usual practise, this image – and the following one – are *not* oriented such that North is up (and East to the right); cf. Fig. 13.1. [© David Malin and Jay Pasachoff, Caltech]

Plate 5

(*Right:*) Again the Crab Nebula, but showing more distinctly its outer confines, thanks to the green line of [OIII]. (*Left:*) Enlargement of the inner
square marked in white, showing the (almost central) pulsar and the half-
luminally expanding pattern of wisps which are brighter in the (approaching)
upper hemisphere. A comparison with Fig. 13.1 shows a rotation of these two
images through some 45° w.r.t. ecliptic coordinates. [Courtesy of Jeff Hester
and Paul Scowen (Arizona State University), and NASA]

Plate 6

A $5' \times 5'$ field from the Orion Nebula – or Orion A, M 42, or NGC 1976 – at a distance of 0.48 kpc, mapped in the composite IR light of {H$_2$ (red), [FeII] (green), 1.25 μm (blue)}. It shows a Hubble flow of age $10^{2.2\pm0.2}$yr, in the shape of outgoing hollow cones – like blown by shrapnel –, centered on the Becklin–Neugebauer Kleinmann–Low IR complex; see Sect. 13.3. [Courtesy of David Allen and Sky and Telescope 86 (1993)]

Plate 7

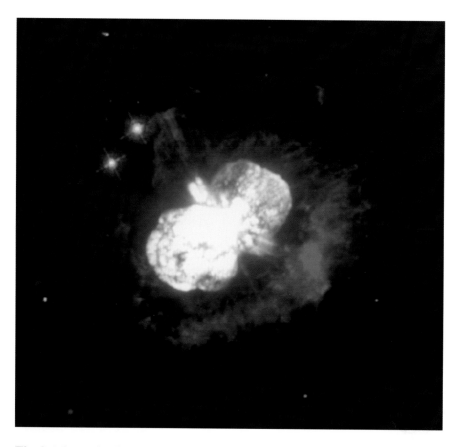

The bright multiple-star system η Carinae hides at the center of the shown Homunculus Nebula, a unique, $\lesssim 10^{2.2}$yr-old emission nebula at a distance of (2.3 ± 0.2) kpc. This 3-colour rgb image, composed of {656, 410, 336}nm, has been adapted from Kerstin Weis [Sterne und Weltraum 6, 436 (2001)]; it is discussed in Sect. 13.6

Plate 8

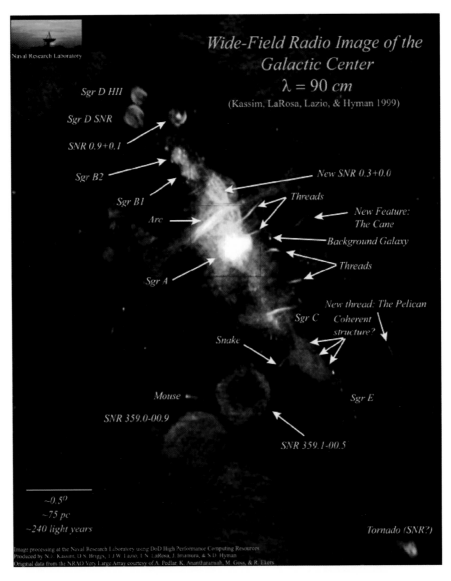

Long-wavelength radio map (90 cm) taken with the VLA, of the innermost few degrees of our Galaxy, centered on the radio source Sgr A and oriented w.r.t. ecliptic coordinates in which the Disk is inclined at 60.2°. All the 'threads', the 'arc', the 'snake', and the 'cane' may be components of the system of coaxial cylinders emphasized in Fig. 13.5. We probably face a micro-quasar activity, see Sect. 13.7. Most of the other, more spherical sources are supernova remnants and HII regions. [Courtesy of Namir E. Kassim and collaborators, and the U.S. Naval Research Lab, NRAO, NSF and ONR; see also Astron. J. 119, 207–240 (2000)]

Plate 9

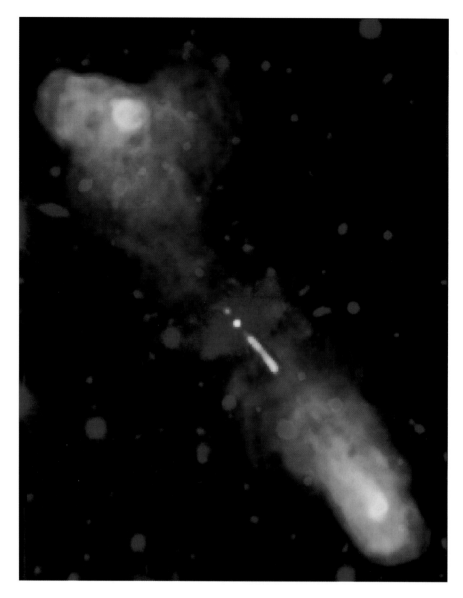

Radio Galaxy 3C 219 = B0917+458, $z = 0.1745$, i.e. at distance $d = cz/H =$ 0.6 Gpc $H_{-17.6}^{-1}$, of projected diameter $\gtrsim 3' \simeq 0.6$ Mpc$(d/0.6$ Gpc$)$, mapped as a radio-optical superposition at $1.4''$ resolution, with radio intensities coded red (near 1.4 GHz) and yellow (near 1.6 GHz), and V-band intensities coded blue; the hot core is white. Note the stretched (type A) morphology, with strong sidedness of the twin jet. [Courtesy of Alan Bridle; see also Astrophys. J. 385, 173 (1992)]

Plate 10

This Spiral Galaxy, NGC 2997 = AAT 17 in Antlia of distance $\gtrsim 10\,\mathrm{Mpc}$,($z = 0.003626$), looks similar to M 83, and may not be very different from our Milky Way. It has been mapped by David Malin with the AAT; the map is $12'$ wide. [© Anglo-Australian Observatory]

Plate 11

Galaxy Cluster Abell 2218, $z = 0.175$, mapped with the HST, of vertical extent $80''$. The center of this cluster projects onto the bright galaxy at the center of the circular arcs, each of which is the tidally distorted image of a background galaxy, via gravitational lensing, of redshift $0.5 \lesssim z_s \lesssim 2.5$. A few galaxies are multiply mapped. The lensing mass, which grows with the impact parameter (measured from the cluster center), can be determined through inverse mapping. [Courtesy of NASA, Andrew Fruchter and the ERO Team (STScI); see also Astrophys. J. 471, 610 (1996)]

Plate 12

On January 15 1996, the HST created this deep 3-colour map of an almost 'empty' field of the sky, exposed for 10^2h net measuring $1.18' \times 1.23'$ [Harry Ferguson], and sampling out to redshifts of $z \lesssim 6$. Several hundred galaxies have been counted on it. Apparently, there are no truly empty patches in the sky. [Courtesy of R. Williams, NASA and STScI]

Plate 13

The Entry of the Animals into Noah's Ark, by Jan Brueghel the Elder (1568–1625). [Courtesy of The J. Paul Getty Museum, Los Angeles]

Plate 14

Index